# Supporting Emergent Multilingual Learners

## in Science

### Grades 7–12

# Supporting Emergent Multilingual Learners in Science

### Grades 7–12

Molly Weinburgh
Cecilia Silva
Kathy Horak Smith

National Science Teachers Association
Arlington, Virginia

National Science Teachers Association

Claire Reinburg, Director
Rachel Ledbetter, Managing Editor
Andrea Silen, Associate Editor
Jennifer Thompson, Associate Editor
Donna Yudkin, Book Acquisitions Manager

ART AND DESIGN
Will Thomas Jr., Director
cover and interior design

PRINTING AND PRODUCTION
Catherine Lorrain, Director

NATIONAL SCIENCE TEACHERS ASSOCIATION
David L. Evans, Executive Director

1840 Wilson Blvd., Arlington, VA 22201
*www.nsta.org/store*
For customer service inquiries, please call 800-277-5300.

*NSTA is committed TO publishing material that promotes the best in inquiry-based science education. However, conditions of actual use may vary, and the safety procedures and practices described in this book are intended to serve only as a guide. Additional precautionary measures may be required. NSTA and the authors do not warrant or represent that the procedures and practices in this book meet any safety code or standard of federal, state, or local regulations. NSTA and the authors disclaim any liability for personal injury or damage to property arising out of or relating to the use of this book, including any of the recommendations, instructions, or materials contained therein.*

Cataloging-in-Publication data for this book and the e-book are available from the Library of Congress.
ISBN: 978-1-68140-481-3
e-ISBN: 978-1-68140-482-0

# Contents

# Dedication

**W**e dedicate this book to all students and teachers who have influenced our thinking. Specifically, for their support, we thank the following:

- Fort Worth Independent School District
- Andrews Institute of Mathematics & Science Education
- Texas Christian University
- Tarleton State University
- City of Granbury
- Our families

# Preface

Instruction using the 5E learning cycle (Bybee at al. 2006) and the *Next Generation Science Standards* (NGSS Lead States 2013) is not enough if emergent multilingual learners (EMLs) are to engage in rich science content and language.[1] Therefore, we offer the 5R Instructional Model (Weinburgh and Silva 2011a, 2011b, 2013; Weinburgh, Silva, and Smith 2014) as a tool for purposeful scaffolding of language in the science classroom. The 5E learning cycle (engage, explore, explain, extend, evaluate) has been a standard in science education for decades. Starting a lesson by eliciting prior knowledge focuses the learner's attention on the task and provides a time for using language. Exploring nature phenomena and collecting data to develop an explanation helps develop not only science content and practices, but also language. This cycle is consistent with the 5R Instructional Model. Each *R* (replace, reveal, repeat, reposition, reload) may be manifested in any of the phases of the learning cycle. Our goal is to produce a resource that middle and secondary teachers can use to learn more about the integration of inquiry-based science and multimodal language instruction for EMLs. As a team engaged in both teaching and researching, we recognize our professional growth as we work to integrate our disciplines: science (Molly), mathematics (Kathy), and language (Cecilia). We invite you to read about our work and join us on this academic journey.

The first four chapters of this book serve as an overview to frame the 5R Instructional Model. Each chapter begins with a scenario describing an instructional event in a science classroom with EMLs. In Chapter 1 (p. 1), we attempt to capture the wide range of EMLs who come together in classrooms where science and language learning are integrated. In Chapter 2 (p. 9), we explore the social language of science and examine meaning-making through the lens of the multimodal language of science: natural language, mathematical expressions, visual representations, and manual-technical operations. In Chapter 3 (p. 21), we focus on science as a context for language learning. We explore inquiry practices of science and conclude with an example of a science lesson. Chapter 4 (p. 35) outlines the 5R Instructional Model and briefly describes the five *R*s.

Chapters 5–9 focus on the five *R*s, and each begins with a scenario that highlights the relevant *R*. A rationale presents how the *R* is situated within the academic context. In Chapters 5–8 (pp. 45, 57, 67, and 75), we discuss and provide examples of the four modes of hybrid language in relation to each *R*. In Chapter 9 (p. 89), we focus on vocabulary as it relates to word meaning in the process of reloading.

Finally, Chapter 10 (p. 97) brings the work full circle. We provide the voices of teachers who have used the 5R Instructional Model as a tool in developing their own science lessons. Thinking of this model as an overlay when using the learning cycle, these teachers express their ability to provide better instruction for EMLs.

---

1   We have chosen to use *emergent multilingual students* to acknowledge the bilingual/multilingual home language practices of the middle and secondary students learning science as well as English in U.S. schools today.

# About the Authors

**Molly Weinburgh**, Andrews Chair of Mathematic and Science Education, is the director of the Andrews Institute of Mathematics and Science Education at Texas Christian University. Her scholarship focuses on academic language acquisition and conceptual understanding in science by emerging multilingual students. She served as co-editor of *Science Teacher Preparation in Content-Based Second Language Acquisition* (Oliveira and Weinburgh 2017) and was coauthor of a chapter outlining the 5R Instructional Model in *The Handbook of Educational Theories* (Weinburgh and Silva 2013).

**Cecilia Silva** is a professor emerita at Texas Christian University. Her research focuses on bilingual education, and she is interested in the integration of content to promote literacy and conceptual development. She began her career as a bilingual teacher. Over the past 30 years, she has worked with teacher preparation programs.

**Kathy Horak Smith** is an associate professor of mathematics at Tarleton State University. Her research has focused on the integration of mathematics and science and communication in the mathematics classroom. She has taught mathematics education courses to preservice and inservice teachers for about 25 years.

# Our Stories

We believe that knowing more about us—the authors—will help situate your learning. The stories that follow are meant to provide you with knowledge of how we came together to think about content integration and multimodality.

Our collaboration began when the English as a Second Language director of an urban school district contacted Cecilia and asked for help with writing a three-week summer enrichment curriculum for emergent multilingual learners (EMLs) enrolled in a newcomer's program. This program, structured as a school within a school, sought to gradually transition EMLs into mainstream classrooms. We began our work in the fall of 2006 and taught the first group of students in the summer of 2007 (Silva et al. 2009).

## Cecilia (Language)

I have always been interested in the junction of language and content instruction. Coming into this partnership, I was ready to make a contribution to the summer program by drawing on my experiences in thematic unit curriculum development. As a bilingual elementary teacher, I had regularly used thematic units to integrate language and content instruction (Kucer, Silva, and Delgado-Larocco 1995). Thematic units provide EMLs with opportunities to use multiple communication systems (e.g., art, movement, language, mathematics) and various disciplines (e.g., literature, social science, science) to explore meaningful ideas. I also saw that, through integration around themes, I could support EMLs at different levels of language and literacy development yet provide all students with the opportunity to engage in significant curriculum conversations.

At the time I entered this project, I was also exploring the works of authors whose research on academic language was making an impact on practices focusing on content instruction for EMLs (e.g., Gee 2004; Gibbons 2006; Scarcella 2003; Schleppegrell 2004). Newer understandings about the nature of academic language in general and the academic language specific to each discipline were reflected in the *PreK–12 English Language Proficiency Standards* published by TESOL International Association (2006). These standards specifically targeted four core content areas—language arts, mathematics, science, and social studies—and aimed to support academic language development for EMLs at different levels of language acquisition. Through the project, I expected to use this growing body of literature to address students' academic language needs.

As a partner in the collaboration, I wanted to implement sheltered instruction (SI) practices (Diaz-Rico and Weed 2010; Rothenberg and Fisher 2007) that were effective in designing lessons to integrate content and language for EMLs. SI lessons support EMLs through a variety of scaffolds that make content comprehensible while promoting language learning. Early in our discussions about the integration of language and content, Molly and Kathy identified multiple areas in which SI lesson design also served inquiry-based instruction. The development of lessons that build on students' backgrounds and prior learning experiences had a long tradition within their respective disciplines. We also anticipated that the hands-on integrated experiences teachers wanted to bring to science and mathematics lessons would naturally serve to enhance meaning-making for EMLs. Although science and mathematics teachers are not expected to plan student interactions to support linguistic development, Molly and Kathy clearly saw how inquiry-based lessons could be modified to promote language learning.

# Our Stories

We also identified areas of conflict between SI and inquiry-based models of teaching, particularly with the Sheltered Instruction Observation Protocol (SIOP) (Echevarría, Vogt, and Short 2000), a model that was gaining popularity at the time. SIOP stressed the front-loading of objectives and vocabulary. In this model, teachers explicitly define, display, and review content and language objectives prior to engaging students in the lesson. Similarly, SIOP emphasizes the preteaching of vocabulary to build background within a lesson. I would learn that this approach conflicts with science inquiry models that stress the need for students to first experience new phenomena through exploration.

## Molly (Science)

Cecilia asked me to help develop curricula for the district summer program. I walked into the first meeting with a knowledge of inquiry-based, activity-oriented science instruction and was well steeped in the *National Science Education Standards* (National Research Council 1996). The five essential features of inquiry were second nature to me, as was the 5E learning cycle recommended by the science education community. In addition, I was beginning to read new science education research examining the multimodal aspects of science communication and meaning-making. However, my education as a biology teacher and 18 years of teaching high school resulted in my knowing almost nothing about teaching language skills.

Because I brought a perspective of inquiry-based instruction to the project, I wanted to begin the unit with a question that students could explore over several days or weeks. This would be a question that could be investigated by changing different variables to determine their effect. I wanted a rich environment with access to physical materials that students could manipulate to find patterns, which would eventually lead them to construct explanations for natural events and behaviors. I also needed students to finish the program with a much deeper conceptual understanding as well as an understanding of the roles that both investigations and models play in science.

An early conversation revealed that mathematics was increasingly important in helping students display and make sense of data, so we invited Kathy to collaborate in the project. Our three-way conversations became professional development for us as we learned more about the role we each played in our own disciplines. Language and literacy ideas that Cecilia recommended made sense in many ways, but I was still concerned. As our team continued to read, discuss, and plan, we sought to weave the three disciplines into a coherent unit. We wanted each discipline to be authentically represented.

We began to focus on two ideas: the position of language—especially the complex vocabulary and discourse of science—and the idea of hybrid language (Lemke 2004). We decided to use science as the cornerstone (content area) of the unit and integrate mathematics and language in authentic ways. This was not easy. However, as we looked at the five essential features of inquiry, we saw that each one requires language and mathematics skills and knowledge (National Research Council 2000). Later, *A Framework for K–12 Science Education* (National Research Council 2012) and the *Next Generation Science Standards* (*NGSS*; NGSS Lead States 2013) provided further notions of the practices of science and how language should be included.

## Kathy (Mathematics)

Cecilia and Molly invited me to join the project because of my background in mathematics, my teaching experience, and my love of children's literature in mathematics. At the time, I was interested in learning how students acquire the vocabulary necessary to be successful in a subject. My questions were not derived from working with EMLs but from 30 years of tutoring students who struggle to pass standardized academic tests. I found that students who were in classrooms where academic vocabulary and academic texts were used were still unsure of how to answer vocabulary-rich contextual problems. Even though I knew many students were struggling because of an inability to comprehend what was being asked, I still did not grasp the complexity of academic language.

My interest in developing vocabulary instruction methods was so great that, even before we began the project, I read several books and attended a nationally recognized language integration workshop. At the workshop, the presenters demonstrated literacy strategies for teaching vocabulary in all subject areas and said that objectives and vocabulary should be introduced at the beginning of the lesson. In mathematics—as in science—this approach is seen as "giving away the punchline" when we want students to develop a concept using a social constructivist view. The workshop presenters also discussed the value of "math talk," journal writing, and trade books but failed to help me understand how these strategies develop academic language. I came away from that workshop with many creative ideas but also with incongruences between teaching vocabulary and teaching mathematics using inquiry and problem-solving techniques. I was left in a state of disequilibrium.

Simply introducing new vocabulary at the beginning of a lesson is not the answer. Academic language must be addressed throughout a lesson so that a connection can be made between vocabulary, content, and context. We therefore begin this book with three assumptions. The first assumption is that the 5E learning cycle used in science education is the cognitive model for instruction. The second is that the three dimensions of the *NGSS* (science and engineering practices, crosscutting concepts, and disciplinary core ideas) are evident in science instruction. And the third is that meaning-making is always socially situated and multimodal. These ideas will be revisited and explored throughout this book.

Our journey has not been fast or smooth. It is also incomplete. We have learned and continue to learn from one another and our colleagues, preservice teachers, in-service teachers, and EMLs. Even as we write, we acknowledge that we continue to learn and change. Perhaps most exciting is that the fields of science, mathematics, and language arts focus on communication and emphasize the value of collaboration and integration.

# Chapter 1

# Introduction

Many middle and secondary science teachers work in schools where English is the primary language of instruction even when it is not the home language of many students. This means teachers often attend conferences to learn new content and language teaching strategies. In the scenario that follows, the teachers are at a National Science Teachers Association (NSTA) conference to learn about the three dimensions of the *Next Generation Science Standards* (*NGSS*; NGSS Lead States 2013): science and engineering practices, crosscutting concepts, and disciplinary core ideas.

*We attended a regional NSTA conference and were impressed by the number of sessions covering all grade levels and topics. What stood out most was a conversation among a small group of teachers as they stopped for a midmorning break. Our attention was drawn to this group because they were discussing something dear to us.*

*One teacher asked, "How can I do all of these things suggested by the presenter of that last session?"*

*"What do you mean?" came the chorus of responses.*

*"The presenter showed us how to engage English language learners in inquiry. It was great! She posed a question and asked us to brainstorm different ways we could find an answer. We were given materials, and we designed our own ways to find an answer. We talked and laughed and redesigned our setup. We gathered data, displayed the data in a graph, and then compared it to everyone else's work. Throughout the session, the presenter*

*engaged us in discussions of how the NGSS stress science for all. She helped us think about how this type of teaching would engage English language learners in rich language and conceptual understanding."*

*"So what's the problem?"*

*"The problem is my district mandates a program that requires me to introduce each science lesson with vocabulary. This seems to be the opposite of inquiry. I kept thinking that if our presenter had given us a vocabulary lesson up front, then we wouldn't need to investigate because the definition would have given us the answers. We would have been less creative and seen fewer ways to come to the same conclusion."*

*"I hear your concern, but I loved the session. I loved that she stressed communication as part of the NGSS practices. Did you notice that she moved beyond vocabulary to thinking about communication? From what we did, I now see science is about communication—not just memorizing words. I'm beginning to think about language differently. I think I can modify the mandated lessons."*

## Rationale

Teachers such as the ones in this scenario influenced this book. Our purpose with this book is to provide a resource for teachers who are developing science lessons for emergent multilingual learners (EMLs). As explained in the preface, we each came to this work from very different vantage points: science, mathematics, and language. Our work is not just about each of our disciplines. Rather, it is about how those disciplines intersect to support meaning-making within the science classroom.

The strength of our collaboration is in the overlap. It's in the way we each approach our content, view the complexity of integrating content with language, and think about teaching EMLs. Working together, we developed and taught lessons that integrate multimodalities within science inquiry, and we anchored our teaching in the research of others (AAAS 1993; Cummins 1996, 2000; Fang, Lamme, and Pringle 2010; Gibbons 2006; Halliday 1993; Kress 2010; Lee 2005; Quinn, Lee, and Valdes 2012; Saul 2004). In addition, we conducted professional development (PD) with other teachers who were interested in supporting EMLs through the overlap of science, language, and mathematics, and we expanded our teaching and PD with our own research (Silva et al. 2012; Weinburgh and Silva 2011a, 2011b, 2012; Weinburgh et al. 2012; Weinburgh et al. 2014).

In this chapter, we provide context for the intersectionality of language and content by developing an understanding of upper-grade science teachers and students and what they bring to U.S. schools. We follow this with an examination of current learning theory before we return to the conflict between inquiry teaching and language instruction depicted in the opening scenario.

## Setting the Context
### *Who are the teachers?*

As a middle or high school science teacher, you were educated to teach content. Your educational background probably did not prepare you to develop skills and strategies for supporting EMLs. You may not have anticipated that you would teach your discipline to students who are also learning English. Your education may have taught you content-specific methods, but it never emphasized the role of academic language within your discipline.

Regardless of your background and current teaching assignment, you are interested in helping EMLs become successful. This means ensuring they can communicate in and about science and can constantly reflect on and clarify their conceptual understandings. In short, you want them to become literate citizens.

### *Who are the children you teach?*

EMLs come from a variety of linguistic, educational, cultural, religious, and socioeconomic backgrounds. Each child, whether he or she is an immigrant, a refugee, or a born U.S. citizen, is a unique mixture of many variables. However, as pointed out by Freeman and Freeman (2009), EMLs who come to middle and secondary school from outside the United States tend to fall into four general categories as determined by English proficiency and educational background. Below, each category is described and an example of a student is provided.

**Educated with some proficiency in English.** Some students come to the United States well educated in their first language and with some English language proficiency. These students know enough English to begin to interact immediately with other students and in their academic setting. They may not know the culture of U.S. schools, but they have an understanding of schooling. They are literate in their first language and already have a sense of the academic language associated with school learning. They need help transferring their knowledge of specific disciplines to the courses they are taking and building their academic language in English.

*Rana, 12 years old, is in sixth grade. She was born in Iraq where her mother was a physician and her father was an engineer. Her family came to the United States when she was 11 years old. She speaks, reads, and writes very well in Arabic. Prior to coming to the United States, she learned some conversational English. Though she is still not fully comfortable with the structure of her new school, she is familiar with the concept of schooling because she attended school in Iraq. She appears to be happy and enjoys reading and soccer.*

**Educated with no proficiency in English.** Other children come to the United States well educated and highly literate in their native languages but with no English language proficiency. These students have a sense of academic language and of the protocols associated with schools.

*Juan is 15 years old and attends a newcomer school. His family moved to the United States from Mexico where his mother was a teacher. Prior to immigrating, Juan attended school and is literate in Spanish. He did not have English instruction while in Mexico. He and his family understand schooling and value education. As a ninth grader, all his subjects are taught in English. He is smart, energetic, and full of humor. He enjoys mathematics, plans to go to college, and wants to be an engineer.*

**Not educated with some proficiency in English.** A third group of children comes to school with some proficiency in English but no schooling. These students, who are somewhat fluent in everyday English, have not developed academic language. Because they have limited schooling experience, they are not familiar with general school routines and expectations.

*Mohamed came to the United States when he was 16 years old. He completed ninth grade and is a year older than most of his classmates. He was born in Somalia and speaks two languages but does not read or write in either of them. His family lived in one of Kenya's refugee camps before gaining entry into the United States. He learned conversational English while at the refugee camp. Although a happy youth, he has periods when he is quiet and reserved. He likes to play mancala, draw, and swim. He wants to be a professional soccer player when he finishes school.*

**Not educated with no proficiency in English.** The last group of students is made up of children with no English language proficiency and no prior schooling. Therefore, schooling and all the unwritten norms associated with it are new to these students. They often are placed in English-only classrooms where they are expected to become familiar with the norms of school, develop literacy for the first time, learn English, and gain content knowledge of various disciplines. These children may struggle in school because they are learning basic English communication skills on top of the culture and norms of schools.

*Mwamba is 13 years old and has just completed eighth grade. He was born in Democratic Republic of the Congo, his family moved to a refugee camp when he was 4 years old, and he came to the United States when he was 11 years old. His family tried to provide him with a basic education, but he has had no formal schooling. Although his history is not completely disclosed, his teachers know that he has seen atrocities to which no child should have been exposed. His mother does not work outside the home; his father works at an assembly plant. He likes to read and do mathematics, and he hopes to one day become a police officer.*

For EMLs who are born in the United States, the issues are similar, though many of these children are like the first of the four previous categories. They have attended school in the United States and are familiar with schooling; however, they may still struggle with learning English.

*Elizbet was born in the United States. Her family moved to Mexico when she was 9 months old and did not return until she was 6 years old and ready to enter school. Because she has not had access to bilingual classes, she only uses English during school hours. Her family, community, and daily life are almost entirely composed of Spanish speakers. In addition, her family returns to Mexico every summer to stay with relatives. Although she is comfortable with her conversational English, she still struggles with academic English.*

For the five students just described, English is both a goal and the language of education. Many teachers can attest to seeing children rapidly become fluent in the everyday language used for interpersonal communication, but studies provide evidence that becoming proficient in academic language takes much longer (Collier 1989; Cummins 1996; Hakuta, Butler, and Witt 2000). Because academic language is complex, Gibbons (2006) suggests that exposing students to the "mainstream classroom without language learning support is an inadequate response to their language development and needs" (p. 5).

## What We Know About Learning and Language

A historical examination of learning theory reveals two main views: (1) the transmission model, in which knowledge is a commodity that can be passed from teacher to learner, and (2) the construction model, in which knowledge is put together by the learner with help from the teacher. The former model, which casts the teacher as a giver of knowledge, was the predominant model in the United States until the advent of the progressive movement in the early 1900s. This model has been compared to banking (Freire 1970) where a teacher deposits information into an empty vault (i.e., the student).

In the transmission model, memorization is highly valued, classrooms are teacher-centered, and lectures are the predominant instructional strategy. Embedded in this model is the assumption that communication is essentially a process of information transfer, and language

is a conduit for this transfer. In other words, teachers talk and students listen. Gibbons (2006) suggests this model also assumes that "language must first be 'learned' before it can be 'used'" (p. 17), which results in the separation of language and content.

The construction model stresses the role of the learner as an active participant in the learning process and emerged with the work of Piaget (1951, 1953) and Vygotsky (1968, 1978). Thus constructivism may be divided into two groups: (1) the personal constructivist view (Piagetian) and (2) the sociocultural constructivist view (Vygotskian). Both have greatly influenced curriculum development in science, mathematics, and literacy.

Piaget (1951, 1953) emphasized active inquiry of the student in the construction of new knowledge. As a developmental psychologist, he focused on the intellectual development of children and how learning occurred. His work suggested that transformative learning occurs as individuals construct knowledge and self-reflect. Learning, an iterative process, requires building new ideas upon what we already know. His work posited stages of development that occur naturally and roughly correspond to the maturation of the child. Also important in his theory is the role of firsthand experiences in helping children build new schema or understanding. For Piaget, language was an outcome rather than a cause of development.

By contrast, Vygotsky (1968, 1978) emphasized the role of social situations and culture on learning. He saw learning as a process of negotiating meanings using the cultural tools of signs, symbols, and language. Vygotsky suggested that there is a discrepancy between the cognitive tasks a learner can do on his or her own and those that can be done with the appropriate help of a more knowledgeable other. He called this the zone of proximal development. Within this zone, learners can actively negotiate and construct knowledge if proper scaffolding is provided. For Vygotsky, language is both a cultural tool for communication and a psychological tool for thinking.

We reject the transmission model and build our work on the constructivist perspective. From our backgrounds in three academic disciplines, we find that both the Piagetian and Vygotskyian schools of thought have played an important role in science, mathematics, and language

instruction. We see how these theories complement each other and have influenced the way we think about the intersection of our disciplines. We believe that children need to be active participants in their own learning as they develop with age. At the same time, we understand the critical role of the adult (or more knowledgeable other) in the learning process.

## Integrating Inquiry-Based Science and Language: Conflicts and Missing Links

Like the science teachers in the earlier scenario, you may also have attended PD designed to help you provide inquiry-based science experiences that follow the 5E learning cycle. Thus, your lessons begin with exploration and are followed by student-initiated explanations. The emphasis in science on inquiry-based lessons is mirrored in mathematics lessons, which stress manipulatives, exploration, and multiple possible ways to solve a problem. Many teachers voice confusion and concern about teaching science as inquiry because they perceive a conflict between inquiry-based lessons and district-mandated programs that require front-loading vocabulary for EMLs.

These concerns are similar to those cited in the research by Settlage, Madsen, and Rustad (2005), who found that teachers struggled with the misalignment between the Sheltered Instruction Observation Protocol and inquiry-based science instruction. Teachers saw the emphasis on communicating language and content objectives and the front-loading of vocabulary to be in direct conflict with their own understandings of inquiry-based science instruction. They suggest that "a fresh conceptualization of the interface between inquiry-based science teaching and sheltered instruction seems necessary" (p. 39). Articulating the appropriate overlap of these has been noted as a constructive step in making high-quality science instruction accessible to EMLs (Fathman and Crowther 2006; Settlage, Madsen, and Rustad 2005).

We purposely set out to find this intersection between inquiry-based science and sheltered instruction as we taught science to EMLs in a large urban school district. As we began to work together, developing an integrated unit based in science with strong literacy and mathematics components, we found that our disciplines did not meld as easily as we first thought they might. Cecilia came from a language background and, like many of you, was comfortable front-loading objectives and vocabulary before engaging students in an integrated science lesson. Molly and Kathy, who were steeped in inquiry-based instruction in science and mathematics, could not reconcile this into their teaching. The two of them, like many of the teachers we encounter, found front-loading specific objectives and vocabulary to be in conflict with national standards on inquiry-based instruction and investigation.

Vocabulary is important; however, it is only one of many complex features that must be acquired for students to be successful in school. Like many content-area teachers, Molly and Kathy were unaware of the demands that academic language places on EMLs and were not prepared to address those demands in the classroom.

Using our strengths, we designed inquiry-based lessons that built on strong science content and the authentic integration of multimodal language in the classroom. We firmly believe that

inquiry-based instruction provides a context in which a deep understanding of both concepts and language can develop. Therefore, we believe the teaching of subject matter content and language should be so integrated that "all content teachers are also teachers of language" (Cummins 1994, p. 42) who "view every content lesson as a language lesson" (Met 1994, p. 161).

## Concluding Remarks

Our work required us to look closely at the intersecting space between content and language and how to integrate instruction to provide authentic language-learning opportunities in the inquiry-based science classroom. Our focus has been on how to help students as they strive to make meaning and communicate their understandings in science. To this end, Chapter 2 (p. 9) examines the multimodal nature of the language used in the science classroom, and Chapter 3 (p. 21) examines inquiry-based science instruction. These chapters are followed by our emerging instructional model for thinking about lesson integration design (Chapters 4–9, pp. 35, 45, 57, 67, 75, and 89). Chapter 10 (p. 97) highlights the voices of teachers who have journeyed with us as we rethink ways to support meaning-making for EMLs in inquiry-based science classrooms.

# Chapter 2

# The Hybrid Language of the Science Classroom

**M**s. Soto emphasizes the crosscutting concept of patterns in both science and mathematics in a crime scene investigation (CSI) unit. In addition, she stresses the science and engineering practices of using mathematics and computational thinking, analyzing and interpreting data, and constructing explanations and designing solutions to communicate information. Ms. Soto gives students the opportunity to engage in all four modes of the hybrid language of science (natural language, mathematical expressions, visual representations, and manual-technical operations) while learning the disciplinary core idea of life science.

*We visited Ms. Soto's biology class, in which emergent multilingual learners (EMLs) were involved in experiences to help them understand human heredity. To anchor the science experiences, she chose to use a CSI unit and had students use blood type, fingerprints, footprints, hair analysis, and DNA as data to solve a crime. Students entered the classroom, put on lab coats and safety glasses, and moved to their seats. We watched as they transformed into scientists.*

*Rather than sitting in straight rows, students were seated at round tables in groups of four, which encouraged more opportunities to express ideas and use equipment. Ms. Soto began by reminding students they have used the word investigation in their science lessons and then asked them to write their own definition of forensic investigation on a sticky note. After a short period, she had the small groups develop a table definition to share with the whole class. The room became a beehive of soft discussions.*

*Building on this opener, Ms. Soto then introduced the footprint as a piece of forensic evidence. Using a large inkpad, students made prints of their own feet and arranged themselves in order*

*from smallest to largest (based on their footprints). Ms. Soto asked students to explain what they observed during this activity, and they noted that the height of the students generally increased with their foot size. This data then went into a T-chart, displaying students' footprint lengths and heights (Figure 2.1). These measurements would later be used to calculate the approximate height of the suspects.*

*This activity led to a discussion about how footprints can help provide information—such as whether a suspect is tall or short—for crime-scene investigators. Ms. Soto suggested that using their own foot and height measurements might show an interesting pattern, and students were asked to graph the data both manually and using a graphing calculator. She stressed that the line of best fit gives a general ratio between foot length and height, and students could then use this ratio to eliminate and narrow down the list of suspects.*

**Figure 2.1.** T-chart showing the ratio of students' heights to footprint lengths

| RATIO OF FOOTLENGTH TO HEIGHT | |
|---|---|
| HEIGHT (X) | FOOTLENGTH (Y) |
| 163 cm | 25 cm |
| 163 cm | 27 cm |
| 148 cm | 21 cm |
| 154 cm | 23 cm |
| 172 cm | 24 cm |
| 170 cm | 22.1 cm |
| 168 cm | 24 cm |
| 150 cm | 21 cm |
| 168 cm | 30.5 cm |
| 162 cm | 27.8 cm |
| 164 cm | 23.6 cm |
| 164 cm | 24 cm |
| 163 cm | 25 cm |
| 171 cm | 26 cm |
| 119 cm | 22 cm |

## Rationale

Students in this scenario are typical of the EMLs in many of your science classrooms. They are expected to engage in the language-intensive, academically rigorous principles outlined by the *Next Generation Science Standards* (*NGSS*; NGSS Lead States 2013) and your state standards. As a teacher implementing these standards, you are aware that instruction has shifted from a focus on scientific facts to the introduction of a framework that requires *all* learners *do* science and use language and mathematical thinking to express scientific meanings (Lee, Miller, and Januszyk 2015). Previously, an English as a Second Language teacher might have been solely responsible for teaching English to EMLs, but now there is the expectation that these students simultaneously learn science concepts and develop scientific language in your classroom. A majority of EMLs spend most of their school day in mainstream classrooms like yours (de Jong, Harper, and Coady 2013; Reeves 2006). Yet, like many teachers, you might feel like you have not received the proper professional development to address their specific linguistic, cultural, and educational needs.

As a teacher supporting EMLs in learning science, you need to become aware of how meaning-making occurs in the classroom so you can create environments where both content and language learning take place. You need to recognize that there are multimodal patterns of communication used in your discipline that go beyond natural language and differ from those that students must use to make meaning in other content areas. As you become aware of the meaning-making features of the science classroom, this book will help you identify opportunities to explicitly model and teach these features within the context of your middle and high school science classrooms.

In this chapter, we discuss ways in which language serves different social purposes. We highlight the social language of science as a unique hybrid that includes four modes, and we conclude with an example that illustrates the components of this hybrid language within the CSI unit presented in the opening scenario.

## Social Language of Science

Lemke (1990) argues that in the classroom, a large part of learning is related to students' ability to use the social language of science. Understanding language from a social perspective is important because it helps support learners as they develop the tools they need to make meaning. When taking this perspective, we go beyond thinking about language as simply speech and writing. In fact, Gee (2002) states,

> There is no such thing as language (e.g., English) or literacy (e.g., reading and writing) in general. People do not learn English. Rather, they learn a specific social language (variety or register of English) fit to certain social purposes and not to others. (p. 162)

To successfully participate in classrooms such as Ms. Soto's, students must learn the language they will need to effectively communicate in science. Furthermore, the *NGSS* specifically state that students must learn to ask questions, construct explanations, engage in argumentation, and communicate information. In addition to learning how to use the *NGSS* practices to communicate, researchers argue that students also need to acquire the identity of science (Gee 2000, 2002; Lemke 1990; NGSS Lead States 2013).

Any individual—a medical doctor, a teacher, an elementary student, a sorority sister, a gang member—develops a social identity within a community and acts in ways that are recognized by members of that particular group. Gee (2002) uses the word *discourse*, which he defines as the "distinctive ways of using (oral and/or written) language, other symbol systems, thinking, believing, valuing, acting, interacting, gesturing, and dressing" (p. 160). For example, students in Ms. Soto's classroom not only learn the natural language of the science classroom but also gain proficiency in the use of particular tools and models, engage in mathematical thinking, develop charts and tables, measure and make estimates, analyze empirical evidence, and use protective gear such as goggles and lab coats, as outlined in the *NGSS*.

We are not yet ready to propose that Ms. Soto's EMLs have fully taken the distinctive socially situated identity of science students, but we are willing to argue that—when engaged in inquiry-based science experiences—they are in the process of developing practices and values that resonate with this identity. As a side note, we remember the moment when Ms. Soto handed out lab coats to her students so they could safely engage in science activities. She introduced the lab coats as appropriate protective gear scientists wear in a lab and compared them to other types of specialized clothing relative to distinct social settings (e.g., shin guards for soccer games, dresses and suits for formal events).

Soon after, students excitedly began to take photos of one another. By the end of the day, they had already displayed their photos on various social media channels. The lab coats took on an even more significant meaning when, two weeks later, the class attended a CSI exhibit at a

local science museum and other museumgoers wanted to know where they had purchased their lab coats. Though they are a seemingly insignificant piece of clothing, the lab coats served to establish a sense of identity that recognized these students as participants in a social community.

Pertinent to this discussion is also the role of language in alienating students in science classrooms (Lemke 1990; Moje et al. 2004). As science teachers, we often contribute to the belief that only the most intelligent can understand the "mystique of science." To bring all children into the science community, we must demystify science and teach students to speak the language of science as a specialized way of communicating about the world. In doing so, we can help students see that scientific conclusions are fallible and science methods are as messy as any other human endeavor. When student identities come into conflict, students often lose their motivation to learn the language styles of the science classroom (Gee 2008). To support students in developing science talk, we need to understand that even though the discourse of science classrooms reflects particular features, it is still grounded in the colloquial language students use to express meaning in everyday situations (Lee, Quinn, and Valdéz 2013). Consequently, we need to provide opportunities for EMLs to use their native languages, the informal English they are in the process of acquiring, and the discourse used in the science classroom.

Gibbons (2015) maintains that as students develop the language of science, they move along a continuum from language that resembles oral conversation to language that resembles written text. The former occurs in contexts where the situation provides extra linguistic cues. For example, a student engaged in a blood-typing investigation in Ms. Soto's class can easily turn to a partner, point to a test tube containing Rh antiserum, and ask that she hand her "that one." In this face-to-face interaction, her lab partner has access to a number of meaning-making cues provided by the context of the conversation. The student can point, use facial gestures, or change the tone of her voice to signal which test tube she wants.

By contrast, language that resembles written text is less situation dependent and places more demands on EMLs. Instead of relying on context for meaning-making signals, students must depend on the language itself to provide enough information to understand the message being expressed. Gibbons argues that there are no absolute boundaries between the two ends of this language continuum. For example, during a class discussion, Ms. Soto uses oral language and a visual representation of a double strand of nucleotides to clarify the role of sugar and phosphates in DNA. Nevertheless, the scientific language she uses resembles written text (like students might find in a textbook) rather than everyday spoken conversation. It's important to remember that, when we socialize students to the language of the science classroom, we must help them gradually move along the continuum.

Scientists act in ways that contribute to the meaning of their disciplines. The particular discourse of the science community is embedded in the natural world of science practices and only makes sense within the social and physical world surrounding this community (Gee 2002; Moje et al. 2004). In addition to listening, speaking, reading, and writing, scientists make use of mathematical calculations, diagrams, and tools to communicate scientific concepts. This means that in science classroom, EMLs must not only acquire social language but also engage in practices that investigate the physical world as they explore natural phenomena.

## Meaning-Making Through a Unique Hybrid Language

When we talk to teachers about supporting meaning-making in the science classroom, they most often identify vocabulary as one of the key challenges their EMLs face. Although vocabulary can be challenging, it is only one of the language elements required for school success. To help students develop the social language of the science classroom, you need to understand what this language entails. Beyond vocabulary, what else is involved in the discourse of the science classroom?

Referring to the social language of science as a unique hybrid, Lemke (1998a, 1998b, 2004) argues that the discourse of science involves four communication modes: natural language, mathematical expressions, visual representations, and manual-technical operations. To conduct science and express the full range of meanings scientists must communicate, they cannot rely on natural language alone. In fact, Lemke (2004) maintains that "scientific literacy is not just the knowledge of scientific concepts and facts; it is the *ability to make meaning conjointly* [emphasis added] with verbal concepts, mathematical relationships, visual representations, and manual-technical operations" (p. 38). A multimodal view of communication in science is supported by other researchers and theoreticians (e.g., Bezemer and Kress 2016; Gibbons 2006; Kress et al. 2001). Kress and colleagues (2001) stress that, even though each mode takes a specialized task, meaning-making in the science classroom is the result of the orchestration of all of the modes together.

Consequently, teachers must be well aware that the choices they make are linked to the options afforded by each mode. Some scientific meanings cannot be communicated with natural language alone. What matters to scientists who are dealing with natural phenomena are descriptions in terms of degree rather than kind (Lemke 2004). Thus, mathematical expressions are needed to quantify precise attributes (e.g., size, shape, speed, rate, density). Similarly, visual representations allow scientists to display details not afforded by other modes of communication. A scaled drawing or a photograph, for example, serves to communicate meanings that are difficult—if not impossible—to communicate using other modes of expression. Finally, within the scientific community, a manual-technical operation—such as pipetting to dispense solutions—is an action that serves to make meaning in the science classroom. In Chapter 3 (p. 21), we will return to ways in which the multimodality of science has been theorized.

Of interest to educators working with EMLs is the meaning-making potential that a broader view of communication affords students (Gibbons 2003, 2006; Molle et al. 2015). For EMLs who cannot access content through English alone, capitalizing on the multimodality of communication in the science classroom increases their opportunity to access similar information while affording them with more opportunities to construct their own meanings (Gibbons 2006; Hand, McDermott, and Prain 2016; Wright 2015).

### *Natural Language*

We already discussed how a large part of acquiring science concepts is related to students' ability to learn the social language of science. In your classroom, there is the expectation that EMLs engage in inquiry and use oral and written language to understand and construct the type of arguments that are at the core of the scientific community. As you work with these students, you support their developing proficiency in understanding the organization and features associated with the text genres of the discipline (de Oliveira and Lan 2014; Freeman and Freeman 2009; Gibbons 2015).

In recent years, linguists have greatly expanded our understanding of the features that encompass academic English (Brisk and Zhang-Wu 2017; DiCerbo et al. 2014; Fang 2006; Fang, Lamme, and Pringle 2010; Freeman and Freeman 2009; Gee 2014; Moje 2007; Schleppegrell 2004; Snow and Uccelli 2009). Despite this increased interest, there still is no consensus about a definition of *academic language*. Some scholars even argue that the diverse perspectives on what constitutes academic language—and the technical language used to describe its characteristics—are off-putting to the nonlinguists engaged in the education of students (Bunch, Shaw, and Geaney 2010).

Our discussion here does not capture the extensive body of literature documenting the linguistic features that characterize the use of natural language within the science classroom. Our intent is to highlight the text genres that align with science discourse and help you support the EMLs in your classroom. Rather than focus on grammar rules, we instead emphasize their communicative functions (de Oliveira 2017; Zhang 2017). This functional view of language and language teaching emerges from research that examines how linguistic features provide students with different options in comprehending and producing meaning as they move along the continuum from oral to written language.

Within this continuum, students develop a wide range of genres. Students entering school settings often have had experiences with personal genres such as narratives; consequently, narratives naturally become the dominant genre within an initial literacy curriculum (Donovan and Smolkin 2001; Pappas 2006; Saul and Dieckman 2005). From a pedagogical perspective, it makes sense to begin the process of literacy instruction by capitalizing on what students bring to the classroom. However, although you might expect to encounter narratives in an English class, this type of text is not common in the science classroom. Scientists not only avoid narration and dramatic accounts but also keep away from dialogue, dramatic action, and other features of personal genres (Lemke 2004). As a result, students in science classrooms are most often expected to engage in *factual* and *analytical* genres (Fang, Lamme, and Pringle 2010; Freeman and Freeman 2009; Gibbons 2015; Schleppegrell 2004; Veel 1997).

Factual genres serve to present facts or events (Table 2.1). Three common text types in this genre are procedures, procedural recounts, and reports. Typically, EMLs develop factual genres when you ask them to plan and carry out investigations. Factual genres make use of the simple present or simple past tense, are discipline specific, and incorporate vocabulary denoting sequence (e.g., *first*, *second*, *next*) or cause and effect (e.g., *if . . . then*, *because*).

Analytical genres, however, require students to investigate phenomena or present arguments that interpret events in a particular way. These include accounts, explanations, and expositions. Students typically engage with analytical genres when they are involved with language tasks: asking questions; analyzing and interpreting data; constructing explanations; engaging in arguments from evidence; and obtaining, evaluating, and communicating information. The linguistic features of analytical genres denote logical relationships, such as cause and effect, comparisons, and conditional statements. We will examine additional linguistic features of the natural language of the science classroom in the chapters that follow as we discuss the 5R Instructional Model (Weinburgh and Silva 2011a, 2011b, 2012; Weinburgh, Silva, and Smith 2014).

**Table 2.1.** Genres and *NGSS* Practices

| FACTUAL GENRES | | | |
|---|---|---|---|
| **Function** | **Text Types** | **Linguistic Features** | ***NGSS* Practices** |
| Present information | Procedure<br><br>Procedural recount<br><br>Report | Present or past tense<br><br>Science-specific vocabulary<br><br>Sequence signal words: *first, second, next, finally*<br><br>Descriptive signal words: *to illustrate, for example, such as* | Planning and carrying out investigations |
| **ANALYTICAL GENRES** | | | |
| **Function** | **Text Types** | **Linguistic Features** | ***NGSS* Practices** |
| Analyze events or argue for certain interpretation of events | Account<br><br>Explanation<br><br>Exposition | Present tense<br><br>Science-specific vocabulary<br><br>Signal words that indicate logical relationships<br>• Cause and effect: *because, due to, as a result*<br>• Compare and contrast: *like, but, similar to, different from, although, however*<br>• Condition: *if . . . then, probably, might*<br>• Taxonomy: *characteristics, for example, including* | Asking questions<br><br>Analyzing and interpreting data<br><br>Constructing explanations<br><br>Engaging in argument from evidence<br><br>Obtaining, evaluating, and communicating information |

*Source:* Adapted from Freeman and Freeman 2009; Gibbons 2015; NGSS Lead States 2013; Schleppegrell 2004; Veel 1997.

## *Mathematical Expressions*

Another component of the hybrid language of science is mathematical expressions. For many years, mathematics was considered the language of science. However, the authors of the *NGSS* recognize it as more complex than that. Specifically, they point out the importance of students analyzing and interpreting data; using mathematics, mathematical expressions, and computational thinking; and communicating information. Not only is mathematical thinking necessary for data analysis, it is also necessary for meaning-making within science. It is our belief that the use of mathematics is needed to do all of the *NGSS* practices.

Mathematics has its origins in natural language but extends it with the use of a set of unique symbols to describe quantitative relations and continuous variation (Cajori 1928; Lemke 2002, 2004). Without such precise ways of describing events, science would not have advanced. Lemke (2004) underscores this by saying, "There is hardly any way in formal verbal language to express subtle differences of degree or ratio. There are no words to distinguish degrees of speed or trajectories of motion" (p. 35). Mathematics is therefore necessary to express both *kind* (typology) and *degree* (topology) (Lemke 2002). For example, in the scenario that opened this chapter, the students in Ms. Soto's class move from stating that the footprint is *large* (typology) to stating its length in terms of a precise measurement of *21 cm* (topology). Having this measurement allows them to set up a ratio, which in turn helps determine the approximate height of the suspect. Students could communicate through natural language by saying, "As the feet get longer, the height of the person gets taller," but a more precise way to communicate a prediction of height is through the mathematical expression $.15 = \frac{x}{y}$, where $x$ = foot length in centimeters and $y$ = height in centimeters.

The unique symbols found in Table 2.2 indicate mathematical expressions, equations, and relationships. Their topological function enables students to be more precise in meaning-making. Without these unique symbols, students would not be able to engage fully in the *NGSS* practices.

**Table 2.2.** Mathematical Symbols and *NGSS* Practices

| Function | Examples of Mathematical Symbols | NGSS Practices |
|---|---|---|
| Numerical | 1, 2, X, L, π, e | • Asking questions<br>• Developing and using models<br>• Planning and carrying out investigations<br>• Analyzing and interpreting data<br>• Using mathematics and computational thinking<br>• Constructing explanations and designing solutions<br>• Engaging in argument from evidence<br>• Obtaining, evaluating, and communicating information |
| Operational | +, -, !, %, √, ∑, ÷ | |
| Grouping | ( ), [ ], { }, **.** (decimal), **,** (comma) | |
| Relational | :, <, >, =, ≠, ≤, ≥, ‖, ≈ | |
| Nominal | ▲, ◊, ○, ∡, ∟, r, C, A, P, Δ | |
| Signal | ⇒, ∀, ∵, ∴, ■ (*If...then, For all, therefore, because, end of proof*) | |

## Visual Representations

In scientific publications, a typical page of printed text contains at least one graphic display and one mathematical expression (Lemke 1998a, 1998b). In these texts, verbal expression only makes sense if it is interpreted in close association with mathematics and visual representations. To make meaning, students in your class must rely on visual systems of representation—in addition to natural language and mathematical expressions. Recognizing the role of visual representations, the *NGSS* calls for students to create, organize, and interpret data through graphical displays and emphasizes that communication within science is done in multiple ways.

In our search to understand visual representation (Coleman, Bradley, and Donovan 2012; Coleman, McTigue, and Smolkin 2011; McTigue and Flowers 2011), we found Moline's (2012) categorical framework to be a useful tool. Similar to natural language and mathematical expressions, visual representations vary in terms of their functions and features. Consequently, some representations serve to express one particular meaning better than others. Moline organizes graphic representation categories by level of complexity, moving from least complex (simple diagrams) to most complex (graphs). Furthermore, he identifies each category in terms of its overall function. Table 2.3 provides examples of Moline's framework.

**Table 2.3.** Visual Representations and *NGSS* Practices

| Function | Type | Examples | *NGSS* Practices |
|---|---|---|---|
| Name parts, enlarge, or show scale | Simple diagrams | Pictures with labels, picture glossaries, scaled diagrams | Developing and using models<br><br>Analyzing and interpreting data<br><br>Constructing explanations<br><br>Engaging in argument from evidence<br><br>Obtaining, evaluating, and communicating information |
| Place information in a particular spatial context | Maps | Bird's-eye view, flow maps, context maps | |
| Reveal and magnify parts that cannot be observed or experienced | Analytic diagrams | Cross section or cutaway, enlargements | |
| Organize a sequence of events | Process diagrams | Timelines, flowcharts, storyboards | |
| Show relationships | Structure diagrams | Venn diagrams, tables, tree diagrams | |
| Measure, rank, and compare | Graphs | Bar graphs, pie charts, spectrum | |
| Organize information so readers easily access information and writers can make their work clearer and more accessible. | Graphic design | Layout, signposts (e.g., headings), typography (e.g., bold, italics) | |

*Source:* Adapted from the work of Moline 2012.

Scientists use visual representations in a variety of ways. Visuals are not redundant in terms of the information they present. Instead, they add information that cannot be entirely expressed through verbal text (Ainsworth, Prain, and Tytler 2011; Anstey and Bull 2006; Lemke 2002; Moline 2012; Prain and Tytler 2013). They display logical relationships that are difficult to describe using language or mathematical expressions alone. Scientists also use visuals when attempting to explain concepts that cannot be observed or experienced directly. Biologists, for example, make use of visuals to represent the replication process of DNA. Similarly, chemists use a figure to represent a carbon ring, which is the fundamental unit for organic chemistry.

### Manual-Technical Operations

The fourth mode of the hybrid language is manual-technical operations. Recently, more interest has been given to thinking about how meaning-making within scientific literacies "emerge from practical actions during joint activity" (Jornet and Roth 2015, p. 379). As stated in the *NGSS*, part of the meaning-making process is being able to manipulate tools and materials within a science investigation context and to analyze and interpret data. We stress that this must be beyond a hands-on or one-time experience. Manual-technical operations may be as simple as setting up an activity to measure water evaporation from an uncovered container or as complex as setting up a distillation apparatus to separate liquids. Even a casual observer would recognize the participants in this mode of the hybrid language as members of a science community.

With each repetition, students develop ease with scientific equipment and practices. Like all other forms of the hybrid language, manual-technical operations must be used within the context of the science classroom. As students' fluidity, automaticity, and competence increase, the focus remains on meaning-making about the underlying science. Thus, students can fully participate and communicate within the science community. Similar to the other modes of the hybrid language, manual-technical operations vary in terms of their function (Table 2.4).

**Table 2.4.** Manual-Technical Operations and *NGSS* Practices

| Function | Examples | NGSS Practices |
|---|---|---|
| Magnify | Hand lens, microscope, telescope | Asking questions |
| Measure | Beaker, triple-beam, ruler, thermometer | Developing and using models<br><br>Planning and carrying out investigations |
| Transport | Pipette, beaker, distillation, dropper, tongs | Analyzing and interpreting data |
| Transfer | Turbine, generator | Using mathematics and computational thinking |
| Contain | Bowl, stream table, round-bottom flask | |
| Safety | Lab coat, fume hood, goggles | Constructing explanations and designing solutions<br><br>Engaging in argument from evidence<br><br>Obtaining, evaluating, and communicating information |

## Crime Scene Investigation: An Example

Often, when discussing the role of meaning-making in the science classroom, teachers remind us that they barely have enough time to teach all of the required subject matter, let alone focus on communication. They also mention that they simply do not feel prepared for this new task. In response, we suggest you begin by considering ways in which disciplinary ideas and the hybrid language intersect in your science lessons. The opening scenario's CSI lesson serves to provide an example.

Ms. Soto began planning by selecting science concepts, skills, and practices most critical for meeting grade-level science standards. In this unit, she wanted students to develop an understanding of heredity as a life science disciplinary core idea. In addition, she systematically identified the hybrid language students would need to understand and express the relevant concepts and processes. For example, she began her lesson by having students orally review their understanding of forensic investigation (natural language). She knew that if students were to understand heredity, they would be required to represent decimal units mathematically (mathematical expressions) and physically manipulate measuring tapes and input data into a graphing calculator (manual-technical operations). They also had to capture data in a way that would allow for later interpretation with labels (natural language) and tables and graphs (visual representation).

In Ms. Soto's class, each meaning-making mode afforded unique options that had to be orchestrated across the hybrid language components. Table 2.5 (p. 20) outlines the components of the lesson as related to the hybrid communication system and the three dimensions of the *NGSS*.

**Table 2.5.** Hybrid Language Identified During the Initial Planning of CSI Unit

| Hybrid Language | Planned Lesson Event | *NGSS* Standards |
|---|---|---|
| Natural language | • Oral explanation of forensic investigation and relation of foot and height measurements<br>• Math/science content vocabulary | **Science and Engineering Practices**<br>• Asking questions<br>• Developing and using models<br>• Planning and carrying out investigations<br>• Analyzing and interpreting data<br>• Using mathematics and computational thinking<br>• Constructing explanations and designing solutions<br>• Engaging in argument from evidence<br>• Obtaining, evaluating, and communicating information<br>**Crosscutting Concepts**<br>• Patterns<br>• Scale, proportion, and quantity<br>**Disciplinary Core Idea**<br>• LS3: Heredity: inheritance and variation of traits |
| Mathematical expressions | • Data collection<br>• Decimals<br>• Fractions | |
| Visual representation | • Ratio of foot length to height table<br>• Foot length to height ratio graph | |
| Manual-technical operations | • Reproduction of foot<br>• Use of graphing calculator | |

## Concluding Remarks

When EMLs enter your classroom, they acquire the discourse of science. As such, they develop the specialized social practices and social language that allows them to recognize and be recognized as members of a particular social community: the science classroom. Social practices include ways of thinking, believing, valuing, and interacting in your classroom. By developing the social language of your classroom, EMLs acquire a hybrid language that includes natural language, mathematical expressions, visual representations, and manual-technical operations. In the following chapter, we examine ways in which you can make use of science inquiry lessons to support the development of the multimodal language of science.

# Chapter 3

# Inquiry-Based Science as a Context for Communication

As the emergent multilingual learner (EML) population continues to grow in many U.S. school districts, teachers are seeking professional development (PD) that addresses content knowledge and language development. The teachers represented in the following scenario attended a PD program for middle and secondary teachers provided by their district. The program stressed the practices of science, and the district science coordinator prioritized science as a venue for language integration.

*Ms. Klopfenstein, a seventh-grade science teacher, joined a group of students as they discussed their lab results and recorded information in their journals. As she looked over the shoulder of one student, she whispered, "I love your drawing. It helps me visualize what you did." She then moved to another group and remarked, "Wow, your measurements are so precise. Your chart is a great way to organize the data so I can easily understand." On to the next group of students and another comment about how quickly they were able to assemble the equipment for the lab activity (a third trial in a series of investigations to determine which variable affected the swing of a pendulum).*

*In the same building, a sixth-grade science class was having a refresher mini-lesson on how to use a stopwatch. Mr. Brown had each student practice stopping and starting the device when he indicated. When students were accurate to within one second, Mr. Brown began to explain the lesson for the day, which required timing how long it takes different materials to absorb water.*

*Several blocks away at a high school, a physics teacher was in the middle of an investigation in which students were examining the work–energy equation. Ms. Rodrigues explained to students that the equation is a model of a causal process, and she reminded them that this idea can be communicated several different ways. She then asked students to brainstorm a list of ways.*

*The murmur of students talking among themselves quickly ensued. Later, students came to the board to display their representations, which included words, mathematical functions, free-body diagrams, pictures, and graphs.*

*Down the hall, students were balancing chemical equations as they prepared for an upcoming investigation. Mr. Huo stressed that using mathematics and computational thinking is essential in science. He reminded students that without mathematical symbols, an explanation of their science investigations would take pages of written descriptions and not be as precise. Without computational thinking, students would not know what to do with their raw data or how to convert it into meaningful information.*

## Rationale

These four teachers are each helping EMLs develop a conceptual understanding of the core ideas within a specific scientific discipline. The students are also engaged in the practices of science as they investigate a pendulum, absorption, work energy, and chemical reactions. These classrooms can easily provide a context for asking questions about real-world phenomena that require students' use of multimodal language to build scientific understanding. By engaging in the practices of science, students can find answers to their questions through the physical and mental exploration of the phenomena. In many cases, EMLs have already experienced the phenomena in some capacity and use their classroom experiences to help build new scientific understandings.

To be successful in science, students must master four major communication/activity domains: doing science (procedures, processes), organizing science (descriptive, narrative), explaining science (sequential, causal), and arguing science (exposition) (Polias 2015). These domains are not confined to speaking, listening, reading, and writing. Although many teachers first think of communication as the purview of natural language, we must think in terms of the multimodal aspects evident in the activity domains. Fully developing the four domains requires students to engage in communication that includes multimodal meaning-making processes and representational choice. Although there are several ways to think about multimodality, we use Lemke's hybrid language as the framework for our work (see Chapter 2, p. 13). Therefore, we stress natural language, mathematical expressions, visual representations, and manual-technical operations used in authentic ways.

In this chapter, we make a case for science as a context that supports the development of the hybrid language. Building on sociocultural theory, we look closely at science as a domain that can act as an apprenticeship for EMLs. This role, which develops as the novice is introduced to the practical application of technical tools and communication modes, provides contextualized learning (Hunter, Laursen, and Seymour 2007). Within this context, we discuss how different modalities fulfill different needs in reasoning and recording scientific inquiry and how learning a new concept cannot be removed from learning how the scientific community represents that concept. We further investigate what is meant by inquiry-based instruction and its importance in developing conceptual understanding and language. We outline how doing science (using the practices of science) is juxtaposed against the elements of the 5R Instructional Model and hybrid language.

The chapter ends with an example that illustrates how a unit on erosion can engage students in developing conceptual knowledge and hybrid language.

## Science as Context

Learning, from a sociocultural constructivist standpoint, is deeply embedded in social contexts and tools (Daniels et al. 2010). This view stresses the active nature of personal and cultural sense-making in which learning is a central and inseparable facet of social practice. It recognizes how learning occurs (process) and the importance of the meaning students make (product). Because science is the study of the natural world, many lessons can be taught by having students interact with phenomena. Active engagement involving the manipulation of materials and open discussion in which ideas are presented result in a context that is uniquely suited to the use of the hybrid language of science and is highly beneficial to its development.

Vygotsky (1968, 1978) introduced the idea of a zone of proximal development, in which the learner is on the cusp of being able to perform at a higher level but cannot do so alone. However, with guidance from a more knowledgeable other (i.e., a teacher or peer), the learner can accomplish the more difficult task (be it manual, linguistic, or cognitive). This type of guidance is often referred to as scaffolding (Gibbons 2002) and is characterized by gradually diminishing the input from the more knowledgeable other and increasing the input (and output) from the learner.

The novice enters the community on the periphery and begins interacting with existing community members. Lave and Wenger (1991) show the importance of legitimate peripheral participation as the "place in which one moves toward more-intensive participation" (p. 36). As the novice becomes enculturated into a community of practice, he or she moves from peripheral participation to full participation. Individuals continue to learn as they move to another, more complex community in which they are again a novice. The community of the science class is an environment in which materials and language are socially mediated and members learn through interactions with others.

Cummins (1981) provides a model to help us understand the interplay of cognition and context. He depicts a continuum of cognitive demands (undemanding to demanding) and contextual support (highly contextualized to not contextualized). Cummins's model shows four ways cognitive load and contextual load can be constructed (Figure 3.1, p. 24). For EMLs, cognitively demanding tasks (i.e., ones they cannot do alone) need to be accompanied with high support (appropriate scaffolding). Science classrooms can provide rich embedded contexts (e.g., hands-on, experiential learning and multimodal support) that support students so the cognitive demands can be increased. As cognitive understanding increases, the need for context-embedded clues decreases.

If a science classroom is designed to require students to talk about their experiences with the natural world, use tools for systematic investigations of that world, and justify their reasoning, then EMLs will develop language and communication skills in an apprenticelike fashion. In this sense, the informal apprenticeship provides an opportunity to work under the guidance of a more knowledgeable other in an authentic environment to learn the tools, methods, and cultures of science over an extended period (Hunter, Laursen, and Seymour 2007; Sadler et al. 2010). An

important aspect is that learning occurs while students participate in tasks that are authentic and relevant both to the cultural practice of science and to the learner. The apprentice gradually becomes more proficient as he or she engages in a task multiple times. In the process, language, habits of mind, practices of science, and content knowledge are acquired and enhanced.

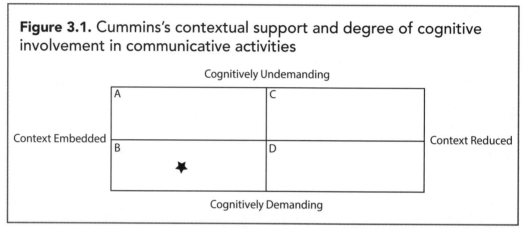

**Figure 3.1.** Cummins's contextual support and degree of cognitive involvement in communicative activities

*Note:* Adapted from *Schooling and Language Minority Students: A Theoretical Framework*, by J. Cummins (p. 12), 1981, Los Angeles: Evaluation, Dissemination and Assessment Center, California State University.

## Multimodality of Science

Learning science concepts and methods involves more than learning to read, write, speak, and listen in English. It involves understanding and conceptually connecting different representational forms or semiotic resources. In recent literature, the term *representational forms* has been used to describe more than one phenomenon. Most often, two different approaches are denoted: multiple representations and multimodal representations. Prain and Waldrip (2006) differentiate the two by saying, "Multiple representation refers to the practice of re-representing the same concept through different forms, including verbal, graphic, and numerical modes, as well as repeated student exposure to the same concept," whereas multimodal representation is used when denoting the "integration in science discourse of different modes to represent scientific reasoning and findings" (p. 1844). Both multiple representations and multimodal representations are important in science.

### *Multiple Representations*

Children need multiple experiences (three to four) with the same concept to establish long-term knowledge (Hattie and Yates 2014). These should be a combination of concrete and nonconcrete experiences, and students should be given the opportunity to produce a variety of representations as they construct and reconstruct their understanding. For example, if students are using a visual representational form to show the change seen in a gel during a DNA analysis, then they should create the visual more than once. To help build conceptual knowledge, students should present a drawing that shows the chamber setup prior to connecting the electrical current, another one in the middle of the experiment, and a third one at the end. This provides multiple times to represent the physical aspects of the equipment and the change that occurs over time.

## *Multimodal Representations*

Murcia (2010) posits that students need to know that science uses multimodalities, and they need to know how to move among them. Each mode of the hybrid language has the ability for meaning-making (Kress 2010). Because modes have different strengths and weaknesses in their clarity, accuracy, and associative meaning, using multimodality provides students with opportunities to revise their work and explore the work from different perspectives (Gunel and Yesildag-Hasancebi 2016). Zhang (2016) points out that "both writing and visuals have the potential to 'sustain' meaning as they can be revisited over and over again. Written language, however, is sequential and takes a logic of succession in time while visual images are non-sequential" (p. 8).

Although several descriptions have been formulated about what is included in multimodality (Jewitt 2017), we use Lemke's four modes of the hybrid language. He explains that scientific concepts are expressed across the modalities, "applying whichever is most appropriate in the moment and freely translating back and forth among them" (1998b, p. 247). He pushes this further by stating that it is in the integration of the modes that the full concept exists. Rather than having translatability within the modes, there is a "complex set of coordinating practices for functionally integrating our uses of them" (1998b, p. 247).

## Inquiry-Based Science

Science educators have advocated for the use of inquiry-based science instruction to learn both the content and practices of science. From the early 1960s, a learning cycle model (Bybee et al. 2006; Karplus and Thier 1967) that stressed exploration before explanation had a major influence on curriculum projects, particularly those funded through the National Science Foundation. Instruction whereby students experience phenomena—either by direct observation or by manipulation of variables to see what happens—to formulate their own ideas has been shown to be an effective way for students to learn (Treagust 2007). This method of inquiry can be particularly effective for EMLs in science content and language development. By manipulating everyday phenomena, students often develop an understanding of concepts even without the language to fully express that understanding (Warren and Rosebery 2008).

However, inquiry-based instruction is also the context for language development as it "shifts the focus of science education from the accumulation of facts and the development of decontextualized science process skills to the provision of experiences that foster the development of scientific knowledge, skills, and habits of mind" (Fang, Lamme, and Pringle 2010, p. 3). With this shift, students are able to develop the processes for generating, validating, and revamping knowledge that scientists use. Scientific knowledge is produced and communicated through hybrid language. Much of this language is unique to each discipline of science and sets the user apart as being a member of a specific group or community. Therefore, students need to be fluent in the discourse of science (as discussed in Chapter 2, p. 9).

Scientists engage in a wide range of activities, depending on the discipline and types of questions within a discipline. All the while, they are developing defensible explanations of the phenomena of the natural world. The context is provided by the investigation, whereas the science-specific forms of language move their thinking forward (Faggella-Luby et al. 2016; Silva, Weinburgh, and Smith 2013). Ultimately, they must use data to develop persuasive arguments around competing explanations (Luft, Bell, and Gess-Newsome 2008). In this process, all four modes of the hybrid language are used.

Although inquiry-based instruction has been encouraged for many years, research indicates that teachers are often confused about their role in providing inquiry-based lessons and about what inquiry looks like in action (Fang, Lamme, and Pringle 2010; Wee et al. 2007). This confusion is not surprising, because inquiry-based science teaching is more than just giving students hands-on materials. It requires new ways of interacting with students and materials. In fact, there is not one specific way that a "good" inquiry lesson should look. This makes having a definitive model and precise examples impossible. As noted in the *Next Generation Science Standards* (*NGSS*; NGSS Lead States 2013), science inquiry is complex, cuts across all disciplines, and is both practice (ways of thinking and acting) and product (content).

The *NGSS* provide a list of eight practices that make up the repertoire of skills scientists use and should be learned and used by students in science classes. With this in mind, it becomes clear that it is not possible to offer a quick answer to the question "What is inquiry?" However, we hope the rest of this chapter provides you with a better understanding of how to develop inquiry-based lessons.

## Inquiry Continuum

We know that many teachers are frightened by the idea of conducting a science lesson as inquiry (Abrahams and Millar 2008; Settlage 2007). This may be because they think they have to jump into full inquiry from the start. Nothing is further from the truth! Science teachers are not expected to conduct every science lesson as full inquiry with students responsible for developing all eight practices. The lesson objectives and availability of phenomena, time, and experience will determine the extent of inquiry used. A continuum of approaches for inquiry teaching is described in several publications (e.g., Fang et al. 2010; National Research Council 2000), and the position along the continuum refers to the amount of the lesson that is teacher-driven versus student-driven.

Researchers using the four variations found in *Inquiry and the National Science Education Standards* (National Research Council 2000) have assigned names to each type. At one end of the continuum is *confirmatory inquiry*, often called cookbook inquiry (level 0 in Table 3.1), in which an activity's main purpose is to confirm an already known fact. In this form of inquiry, students follow the teacher's instruction for all parts of the investigation. Often, the investigation is seeking a specific answer to a question. Although little inquiry is involved—as you do not want your students to experience *only* this type of investigation—a confirmatory inquiry can help students increase their skills in manipulating scientific materials or confirming an established protocol.

At the other end of the continuum is *open-ended* or *full inquiry* (level 3 in Table 3.1), in which students are responsible for all aspects of the investigation. Full inquiry requires students to develop a scientifically oriented question, carry through with a self-proposed investigation, and articulate a defensible explanation to the original question. Science fair investigations are a familiar form of this approach. Full inquiry requires considerable practice and often does not work well the first time that the teacher and the students engage in it. Keep in mind that it is not appropriate for all science teaching.

Between these two extremes on the continuum are *structured inquiry* (level 1 in Table 3.1) and *guided inquiry* (level 2 in Table 3.1). In a structured inquiry, students are given the question, procedures, and relevant information but must display the data in their own way and argue their own conclusions. Usually, this inquiry has a small range of possible conclusions rather than one definitive answer. It allows students to discuss some variation in their interpretations but focuses the possible interpretations. By contrast, in a guided inquiry, students are given a question to answer but must develop the method and draw conclusions. This form of inquiry lesson provides all students with the same starting point but gives much more freedom in how they will pursue an answer, display their results, and argue their conclusions.

**Table 3.1.** Levels of Inquiry With Descriptions and Examples

| Component | Level 0 | Level 1 | Level 2 | Level 3 |
|---|---|---|---|---|
| **Title** | **Confirmatory** No inquiry | **Structured** Low level of inquiry | **Guided** Moderate level of inquiry | **Open** High level of inquiry |
| **Description** | The teacher provides all essential features of the laboratory experience, so no inquiry occurs. | The teacher provides all features, but the analysis of data is completed by students. | The teacher supplies the question, and students provide the remaining features. Students develop the plan, interpret the data, and explain the results. | The students provide all components. Students generate the question, develop the plan, interpret the data, and explain the results. |

*Continued*

**Table 3.1.** (*continued*)

| Component | Level 0 | Level 1 | Level 2 | Level 3 |
|---|---|---|---|---|
| **You use when ...** | Teaching a specific skill or how to use a particular piece of equipment. | Teaching specific content objectives and when making comparisons within an investigation. | Teaching specific content objectives and when multiple ways of providing the answer are acceptable. | Teaching process objectives rather than specific content. Students work with a concept long enough to ask their own questions to extend current classroom investigations. |

## Types of Investigations

Investigations can be classified by the questions that are being asked: *descriptive, comparative,* and *experimental* (West 2010). Descriptive studies are the predominant type used in lower grades and are also widely used by many scientists. These investigations do not change variables but require students to observe and carefully record the current state of the phenomenon. An example would be watching a chrysalis as a butterfly emerges or a seed as it germinates. This type of investigation does have a research question (e.g., What happens as the seed germinates?), but no hypothesis is put forth because the investigation is always exploratory. Descriptive investigations still require measuring skills, data recording, drawing and labeling, and writing.

The second type is the comparative study, which generally looks for a correlation between variables. This type compares two or more groups or events on one variable. This can be seen when testing different conditions on the same phenomenon. The researcher has a question and can suggest a hypothesis. The manipulated variable is identified and placed on the $X$-axis of a graph, and the responding variable is observed and placed on the $Y$-axis. The graph then shows the relationship of $X$ to $Y$. A typical example for middle school students is investigating the variable effects caused by the rate of a swinging pendulum. The manipulated variable can be the length of string, added weight, or thickness of string. The responding variable is the rate (period) of swing. Comparative studies stress the idea of a fair test and the importance of changing only one variable at a time.

The last type of investigation is the experimental study, in which the research question is trying to show causation. This requires both a control and an experimental setup. The variables tested are always compared back to the control (or undisturbed) setup. An example would be investigating the effect a designated medicine has on a given population. Half of the participants would receive the medication (experimental), and half would receive a placebo (control). If there is a difference between the control and experimental groups, causation can be assumed.

All three types of investigations require students to keep good notes; therefore, a journal is useful. Students use mathematical and computational thinking as data are collected and analyzed. In addition, keeping a science journal provides students with the opportunity to produce informational text and use the hybrid language of science.

## Practices of Science

At first, you may think the practices of science are only about the manipulation of materials. Actually, doing science includes both manual and cognitive components. In preparing for a science unit, you need to think about what you want students to learn and what practices are necessary for the unit to qualify as inquiry-based. In an attempt to help teachers understand inquiry, *Inquiry and the National Science Education Standards* (National Research Council 2000) outlined five essential features: learning to ask scientific questions, gather evidence to use in answering the questions, formulate explanations using the evidence, evaluate alternative conclusions, and communicate findings. Very few lessons fully engage students in all five features, but over the course of study, students should have the opportunity to develop their skills in all features.

With the introduction of the *NGSS*, we refined our own understanding of inquiry as outlined in the practices of science, crosscutting concepts, and disciplinary core ideas. The practices include the aforementioned five essential features with three additional categories: developing models, analyzing and interpreting data, and using mathematics and computational thinking. As you help move students toward the construction of their own knowledge and development of specific science-related skills, lessons will vary in the amount of structure you provide for each feature, the complexity of the tools needed, and the amount of responsibility students take for each feature. Students should eventually be able to initiate each component on their own by using the eight practices outlined by the *NGSS*.

In many ways, thinking about the practices of science aligns with our 5R Instructional Model as they give more emphasis to the hybrid language. The 5R Instructional Model, as you will read in Chapter 4 (p. 35), is a nonlinear representation for building language and cognition into the units you plan. The model is grounded in the assumptions that hybrid language (1) emerges in context from the experiences children have with natural phenomena, (2) builds on the hybrid language they already know, and (3) pushes them to use new scientific discourse.

The first column of Table 3.2 (p. 30) lists the eight practices of science. The second column provides descriptions of how each feature may be enacted in a classroom with mathematics and language. The third column gives short examples of which component of the 5R Instructional Model the practice may help provide. The last column points out the modes of the hybrid language that are most often associated with each practice. Chapter 4 explains the 5R Instructional Model more fully and shows its relationship to the 5E learning cycle.

These practices cannot be developed in classrooms where the teacher does most of the talking and is the main source of ideas. Students, especially EMLs, need to engage in each of the practices with the corresponding components of the hybrid language and the opportunity to replace, reveal, repeat, reposition, and reload.

**Table 3.2.** Enacting the Practices of Science With Examples From the 5R Instructional Model and Hybrid Language

| Practice | Enactment in Class With the Addition of Mathematics and Language | Elements of the 5R Instructional Model | Elements of Hybrid Language |
|---|---|---|---|
| Asking questions | These questions should be worded in a way that makes it possible for students to design an investigation. Scientific questions aim to understand origins, causes, and processes. Students should ask questions such as "What is happening?" "What is the purpose for this structure/event?" and "How does this work?" Questions that begin with *why* should be avoided because they are not scientifically oriented and cannot be investigated. In articulating and writing the question as part of a journal entry, students engage in communication skills. | *Repeated* opportunities to develop the language of a good question.<br>*Reposition* as more precise language is used. | Natural language, mathematical expressions |
| Developing and using models | Models may be a static representation of a natural phenomenon that is used to visualize aspects of the phenomenon, such as a globe. They may also be dynamic and used to manipulate variables so students can see cause and effect, such as a stream table. | *Repeated* opportunities to use language and materials.<br>*Replacement* of nonscience language and concepts.<br>*Revealing* new language and equipment.<br>*Reloading* as a review of concepts and vocabulary. | Natural language, mathematical expressions, manual-technical operations |
| Planning and carrying out investigations | An investigation may be conducted in the classroom or outside. This requires students to talk to one another to determine what data are needed, how to record the data, and the manipulated and responding variables. As the investigation is carried out, measurements or the occurrence of the mathematical modeling may occur. | *Repeated* opportunities to use language and materials.<br>*Replacement* of nonscience language and concepts.<br>*Revealing* new language and equipment.<br>*Reloading* as a review of concepts and vocabulary. | Natural language, mathematical expressions, visual representations, manual-technical operations |
| Analyzing and interpreting data | Tools used to analyze data include tabulation, graphical interpretation, statistical manipulation, and visual representation that help determine patterns. Interpretation may result in providing several explanations. | *Repeated* opportunities to use language and materials.<br>*Replacement* of nonscience language and concepts.<br>*Reposition* as more precise language is used.<br>*Revealing* new language and equipment.<br>*Reloading* as a review of concepts and vocabulary. | Natural language, mathematical expressions, visual representations, manual-technical operations |

*Continued*

**Table 3.2.** (*continued*)

| Practice | Enactment in Class With the Addition of Mathematics and Language | Elements of the 5R Instructional Model | Elements of Hybrid Language |
|---|---|---|---|
| Using mathematics and computational thinking | To frame reasonable explanations, students use data from primary or secondary sources. Logic, reasoning, and critical analysis are used rather than personal beliefs, myths, and religious values. Evidence is often mathematical and presented through natural language, visual representations, and mathematical symbols. | *Repeated* opportunities to use language and materials. *Replacement* of nonscience language and concepts. *Revealing* new language and equipment. *Reloading* as a review of concepts and vocabulary. | Natural language, mathematical expressions, visual representations, manual-technical operations |
| Constructing explanations | Evidence is gained in the form of primary and secondary sources. Primary sources are the direct results of student investigations. Secondary sources are often books, the internet, or multimedia resources that require literacy skills. Students determine what evidence is needed to answer the question, how to collect the data, and in what form the data should be displayed. The accuracy of data depends on multiple iterations and careful measurements and is subject to verification. Data may need manipulations, such as deriving the mean or calculating velocity. Data are often presented as numbers and units of measures, using mathematical formulas, graphs, charts, and tables. Explanations may be written, verbal, or pictorial. | *Repeated* opportunities to use language and materials. *Reposition* as more precise language is used. | Natural language, mathematical expressions, visual representations, manual-technical operations |
| Engaging in argument from evidence | Other ways of interpreting similar data are shared through discussions and writings. Students develop a sense that science is tentative, and explanations can become refined or altered. Skills of debate help students with the logic of argumentation. | *Repeated* opportunities to use language and materials. *Reposition* as more precise language is used. | Natural language, mathematical expressions, visual representations, manual-technical operations |
| Obtaining, evaluating, and communicating information | Multiple ways of communicating are encouraged. These include mathematical formulas, graphs, charts, drawings, pictures, and written and oral language. | *Repeated* opportunities to use language and materials. *Replacement* of nonscience language and concepts. *Revealing* new language and equipment. *Reloading* as a review of concepts and vocabulary. | Natural language, mathematical expressions, visual representations, manual-technical operations |

## Erosion: An Example

An example of level 2 inquiry-based instruction can be seen with sixth-grade EMLs as they investigate the concept of erosion. On the first day of a unit, Mrs. Malo posed a big question that takes multiple investigations over several days or weeks to answer. After telling students about her backyard—described as having a slope, sandy soil, and no vegetation—Mrs. Malo posed this question: "What will happen to my yard if we have strong winds as predicted by the weather forecaster?"

Students first decided how to represent the yard by creating a model with a stream table and sand. Although there are infinite ways to make a model of the yard, upon consensus, the same procedure was used for all subsequent trials. Mrs. Malo introduced the idea of a fair test in science because students needed to know they were only allowed to manipulate one variable at a time.

Next, students discussed ways to create the wind, coming to a consensus on one way to be used by all groups. Lastly, students conducted the investigation, collected data, displayed that data, formulated an answer to the original question, and communicated their results (see Figures 3.2, 3.3, and 3.4).

To see patterns across the student groups, Mrs. Malo scaffolded instruction by modeling how to make a group data chart. By creating a group chart, students were able to publicly display their data, and the class was able to discuss patterns that materialized. In addition, outlier data were examined for possible reasons. Mrs. Malo helped students see that although their answers differed somewhat, they all concluded that sand is moved by a natural force (i.e., erosion will occur).

**Figure 3.2.** Student using a straw to test the manipulated variable of wind on the yard

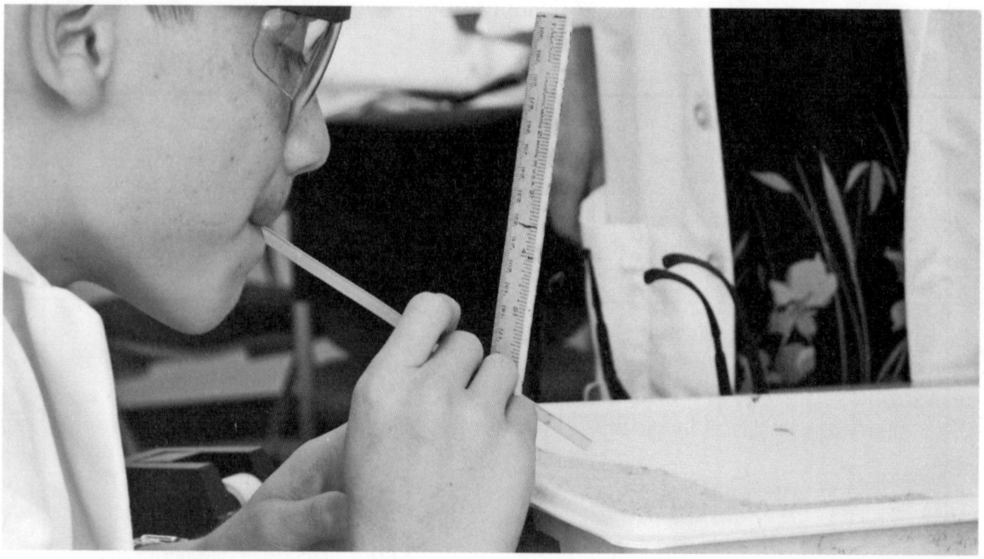

**Figure 3.3.** Student display of data from each table to see if there are patterns that can lead to generalizations

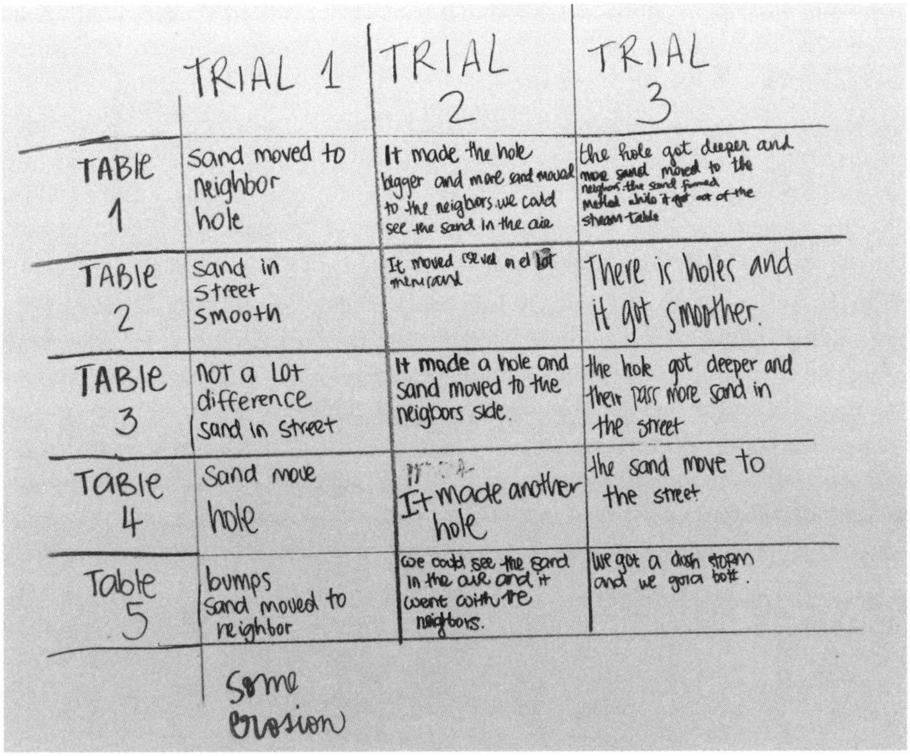

| | TRIAL 1 | TRIAL 2 | TRIAL 3 |
|---|---|---|---|
| TABLE 1 | Sand moved to neighbor hole | It made the hole bigger and more sand moved to the neigbors. we cold see the sand in the air | the hole got deeper and more sand moved to the neigbors. the sand formed metal while it got out of the stream table |
| TABLE 2 | Sand in street smooth | It moved several a lot menu ranu | There is holes and it got smoother. |
| TABLE 3 | not a lot difference sand in street | It made a hole and sand moved to the neigbors side. | the hole got deeper and their pass more sand in the street |
| TABLE 4 | Sand move hole | It made another hole | the sand move to the street |
| Table 5 | bumps sand moved to neighbor | we could see the sand in the air and it went with the neighbors. | we got a dash storm and we gora bott . |
| | Some erosion | | |

**Figure 3.4.** Example of how the manipulated variable of hard rain resulted in a deep gulley, a wide alluvial fan, and runoff in the street

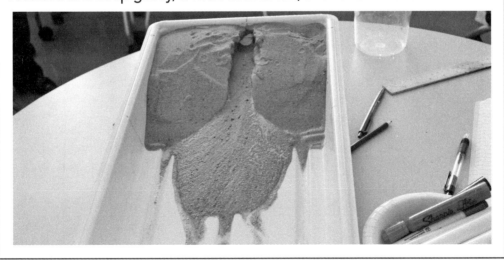

The investigation was expanded when students were asked, "What will happen to the yard if we have a gentle rain as predicted by the weather forecaster?" The process of discussing ways to use the model to answer this question and of conducting an investigation was repeated. This time, students were expected to create a group data chart without their teacher's help. This investigation was followed by one last question: "What will happen to the yard if we have a hard rain as predicted by the weather forecaster?"

Mrs. Malo may have asked the original question, but all other aspects of the investigation were student-driven (level 2, guided inquiry). Throughout each unit and over an entire year, investigations in your class should move back and forth between the levels of inquiry. Over the course of several units, students engage in having different amounts of control over the eight practices of science. The result is that students know how to do all parts of the scientific practices.

For EMLs, structured (level 1) and guided (level 2) inquiry provide wonderful opportunities for input and output as they use the hybrid language of science. Natural language is used in discussing and writing about what they propose as a way to answer questions given to them and again as they implement the plan. Natural language is enhanced and strengthened as they use mathematical expressions in collecting, calculating, and representing data. Visual representations are often used to supplement or enrich natural language, and the very act of working with tools and props of science provides a manual-technical component.

## Concluding Remarks

Because many science units have the potential to be taught as inquiry and can have students investigating real phenomena, science provides a perfect context for developing the hybrid language. Within many lessons, EMLs need to use natural language, mathematical expressions, and visual representations as both input and output. In addition, hands-on experiences (physical manipulation of materials and tools) support the meaning-making process.

As a teacher, you have the responsibility to design lessons that will provide EMLs with opportunities to develop a conceptual understanding of science, skills, the practices of science, and the discourse of science. In the next chapter, we will describe the 5R Instructional Model, which is a tool for instructional design.

# Chapter 4

# The 5R Instructional Model for Science Instruction for Emerging Multilinguals

Mrs. Jones teaches science in a middle school where most of the students are emergent multilingual learners (EMLs). She emphasizes the crosscutting concept of cause and effect in an environmental water unit. In addition, she focuses on the *Next Generation Science Standards* (*NGSS*; NGSS Lead States 2013) practices of using mathematics and computational thinking, analyzing and interpreting data, and constructing explanations to communicate information. She gives her middle school students the opportunity to engage in all four modes of the hybrid language of science while learning the disciplinary core idea of earth and space science. The lesson follows a 5E learning cycle format as students explore real phenomena and collect data to create an explanation.

*Students collected beakers of water from a nearby creek and were using a Secchi disc to test their samples. Each small group measured the distance the Secchi disc could be lowered until it was no longer visible. They added both the depth in centimeters and the JTU (Jackson Turbidity Unit) on a recording sheet (Figure 4.1). After giving students an opportunity to discuss the investigation at their tables, Mrs. Jones asked one child to respond.*

*"Liam, can you tell what you and your partner observed when you used the Secchi disc?"*

**Figure 4.1.** Examples of a student journal, Secchi disc, and recording sheet

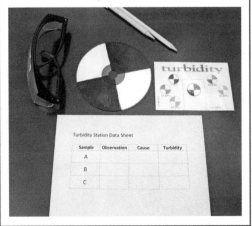

*"We couldn't see it. The water was cloudy."*

*"Interesting. Did any other groups have cloudy water?" All of the groups agreed that they observed cloudy water. "Do you know any other names for 'cloudy water'?" There was a long pause for responses, and then several students offered alternative words, such as* dirty, muddy, and foggy.

*Mrs. Jones acknowledged their suggestions and stated, "Scientists have a word for cloudy, muddy water. They say the water is* turbid. *The more turbid the water, the less you can see through it. A scientist would also say that the investigation you just did allowed you to check the turbidity of the water." She wrote* turbid/turbidity *on a paper strip and placed it on the classroom's word wall.*

*"Turbid is an adjective used to tell about or describe one characteristic of water. Turbidity is a noun. Now that we have this new term, use it in your journals when you are describing cloudy water."*

*Later in the lesson, when a student had the opportunity to recap what she and her group observed during the investigation, she proudly stated, "We saw a turbid."*

*Mrs. Jones replied, "I love that you are using your new word. Let's think about that word.* Turbid *is not a noun; it is an adjective and tells us a characteristic of the water. So you would say, 'We observed that our water was turbid' or 'We observed turbid water.' Do you hear the difference in the way you said it and the way I said it?"*

*Mrs. Jones asked each group to tell her about the investigation, helping students negotiate the concept and ways to communicate what they did and learned. She then informed students that one more test was needed to fully determine the condition of their water samples. She showed students a new piece of equipment and explained, "This is a digital chlorine analyzer. The digital chlorine analyzer is used to test the amount of dissolved chlorine in your water sample."*

*Because students did not have a common name for the equipment, she revealed the name, added it to the word wall, and demonstrated how it works. Students used the equipment to test for chlorine and then recorded the parts per million of dissolved chlorine on their data chart.*

*At the beginning of the next lesson, Mrs. Jones used the word wall to reload the new terms (Secchi disc, turbidity, chlorine, digital chlorine analyzer) and the related scientific concepts. Students were to select a term and discuss it with a partner. Once both partners had talked about the term, they switched partners and repeated the process with a new term. After this short reminder of the previous lesson, the new lesson began.*

## Rationale

As you read about Mrs. Jones's class, you may have recognized several scaffolds you are already using with your own students. You now know that she and her students engaged in multi-modal meaning-making using the hybrid language of science, and she started the unit with the manual-technical operation of collecting and testing water from a stream near the school. This helped situate the lesson and provide the materials with which students would work. The students continued manual-technical operations as they submerged the Secchi disc into their water samples and measured the depth.

Mrs. Jones used natural language in ways that support learning by providing new language as she **replaced** some terms students already used and **revealed** new words. She made use of mathematical expressions as students made measurements and described the quantity of dissolved particles. As students recorded data and wrote lab reports, they were encouraged to draw representations of the creek and tools used to collect and analyze the data.

Mrs. Jones asked students to discuss with partners, wrote new terms on the board, and held up scientific equipment for students to see. Throughout the lesson, she **repeated** the use of terms in context while she helped students develop sentence structures that resemble the way scientists talk to one another and write in their journals. She also **repeated** the practice of using the Secchi disc. She helped students develop the language of science by **repositioning** her statements so students heard language more like scientists use it. Finally, she used her word wall as a way to capture important words and then to come back to **reload** them in later lessons.

Some of you may find yourself in a similar position to the one in which we found ourselves—wanting a tool for preparing lessons that engage students the way Mrs. Jones was able to do. You know that engaging science lessons are inquiry-based, which require students to seek answers to interesting questions. Learning in the science classroom involves acquiring a new social language to learn and communicate (and even think about) ideas and procedures. You are also aware that children develop their informal discourse within the communities in which they are first socialized. They then enter school, where they acquire new styles of language and new ways of acting, feeling, and thinking. All children, but especially EMLs, must learn the language of academic disciplines they encounter to succeed in school.

In our effort to support inquiry-based lessons and hybrid language learning, we developed the 5R Instructional Model. We envision this model as an instructional tool for designing

lessons that provide scaffolds to support EMLs in science classes. The principles undergirding the 5R Instructional Model, as we discussed earlier (p. 29), are that hybrid language (1) emerges in context from the experiences students have as they interact with natural phenomena, (2) builds on hybrid language students already know and can use, and (3) pushes students to use new scientific discourse as a way of making meaning in the science inquiry classroom.

As we discuss the 5R Instructional Model, we stress that the model is *not* linear, and it varies with each use. It is a tool that can help frame your planning, including how to scaffold instruction to meet the needs of EMLs within the context of inquiry-based science lessons.

In this chapter, we build on the notion that EMLs must develop the hybrid language you read about in Chapter 2 (p. 9) to be successful in school. We revisit and extend the concept of scaffolding (first discussed on page 23), and we end the chapter by introducing the components of the 5R Instructional Model.

## Scaffolding

Though natural language is often considered the dominant mode of communication in the science classroom, students also need to be supported and scaffolded in comprehending and constructing meaning through mathematical expressions, visual representations, and manual-technical operations (Kress et al. 2001; Lemke 2004; Molle 2015). Scaffolding, as we discussed in previous chapters, refers to the type of support that science teachers offer all students—but especially EMLs—as they develop science concepts, practices, and hybrid language in the classroom. Going back to Vygotsky's (1968, 1978) concept of the zone of proximal development (as described on page 23), we now focus on ways in which EMLs collaborate with more knowledgeable others (e.g., classroom teachers and capable peers) to perform tasks they cannot yet accomplish on their own, given their level of language development.

A decision to scaffold hybrid language within the science classroom reflects an understanding that if they are to *do* science, EMLs must develop the full range of scientific language. As you plan science lessons, you need to consider scaffolding tasks that explicitly develop scientific discourse while developing scientific concepts. Here we again remind you that we are not talking about teaching grammar or vocabulary in isolation. Instead, we need to give more explicit attention to teaching students how to use multimodal language as a meaning-making tool within the context of science lessons. An explicit focus on the hybrid language of science provides EMLs with additional ways of making sense of the new knowledge and concepts they typically encounter in the classroom.

For example, a technique we often use to scaffold hybrid language is the creation of data tables. Data tables allow students to use natural language and mathematical expressions to display data organization visually. The teacher can help create the first data table and point out the importance of each column and row. Students are then able to use that as a model for future data tables. Teachers should also model how to make a quick sketch of equipment or results and remind students that making a sketch is beneficial as a way to remember what they used or what happened. By doing so, students have a model for capturing information visually.

A critical characteristic of scaffolding is that the more knowledgeable other adapts the level of language support offered so students, ultimately, can perform tasks independently. As we scaffold, we are aware that even though we are acting in the present, we are envisioning the future in terms of students' potential development. Often, classrooms accomplish this with strategies such as "I do, we do, you do," in which the teacher models before students work in small groups and then finally alone.

Although the literature on scaffolding language and learning is extensive, the work of Hammond and Gibbons (Gibbons 2009, 2015; Hammond 2014; Hammond and Gibbons 2005), which conceptualizes scaffolding in terms of two levels of student support, has particularly informed our practice. Hammond and Gibbons use the terms *macro-scaffolding* and *micro-scaffolding*. Macro-scaffolding refers to the planning decisions teachers make as they include language tasks in the curriculum. Micro-scaffolding, by contrast, refers to the collaborative work that occurs between student and teacher within the context of the lesson. Whether scaffolding is macro (planned) or micro (spontaneous), it is a way of "*supporting-up* such students, rather than *dumbing-down* the curriculum" (Hammond and Gibbons 2005, p. 6).

## Macro-Scaffolding

Macro-scaffolding includes many of the planning features you might expect in any class where there is good teaching. Thus, we ask you to consider the critical features of macro-scaffolding through the lens of a science educator who understands that teachers must make the hybrid language explicit. Next, we identify and discuss some of the features considered essential to the planning of macro-scaffolding in the classroom.

**Background knowledge.** Even novice teachers can identify background knowledge as the backbone of lesson planning, because it takes into consideration EMLs' prior experiences with science content and their level of language proficiency. Students are only capable of building new concepts and language on previous knowledge and understandings (Vygotsky 1978; Walqui 2006). Therefore, to help students build new knowledge and ultimately develop the discourse of science, you must start where they are.

**Metacognitive awareness.** As a science teacher, you are aware of the benefits of supporting metacognition—thinking about thinking—in your classroom. This involves scaffolding learners to become aware of their own thinking processes as they explore new science practices. In effective language classrooms, teachers explicitly teach skills that help students become aware of how and when to use appropriate cognitive approaches to accomplish any given task. Teachers use strategies such as open-ended questioning, reflective journals, and think-alouds to scaffold students' ability to monitor and reflect on their own learning. This type of metacognitive awareness is critical for second language learners (Chamot 2005; Chamot and O'Malley 1994; Macaro 2006; Oxford 1990) because it helps solidify their learning.

**Metalinguistic awareness.** Metalinguistic awareness refers to the ability to use language to reflect on and explain how language functions. It serves to make language visible to EMLs. What this means in the science classroom is that teachers purposefully plan opportunities to support students' understanding of how different language features function when constructing meaning. Students must have both understanding of the function and the language to communicate this knowledge. For example, when discussing ways to support students' construction of science arguments, Hand (2008) proposes that students must not only learn to use the language of science but also learn the language of the argument (e.g., claim, evidence) and its function. On a similar note, Serafini (2011) argues that as students become aware of the various functions of a visual representation, they develop the vocabulary to name and describe its various elements (e.g., line, pattern, shape). Developing the metalanguage for describing the features of the hybrid language supports students' ability to understand, reflect on, and use the language. When we name something, students are more likely to notice, recognize, and use it within the context of a science experience (Gibbons 2009).

**Interaction.** As teachers plan language tasks in the classroom, they create opportunities for comprehensible output (Swain 1985, 2005). Comprehensible output refers to the ways in which teachers support EMLs in using the language of the science classroom. This again involves the intentional planning of activities in which students use the hybrid language to communicate science understandings with a partner, in small groups, and as part of whole-class discussions. Science teachers use a variety of cooperating learning strategies with which they are already familiar to scaffold language output for EMLs (Kagan 1995; Slavin 1995). These strategies are particularly beneficial as they are structured so team members are accountable for specific tasks needed to accomplish the group goal.

## Micro-Scaffolding

Micro-scaffolding is the spontaneous language support given to EMLs when they need to accomplish a task they cannot do on their own. This type of scaffolding is offered within the context of a lesson, as skilled teachers observe learners and identify a teachable moment. Cazden (1988) refers to these teachable moments as "in-flight interactions." Even the best-laid lesson

plans can "go awry, because no lesson is under the teacher's unilateral control; instead, teacher behavior and student behavior reciprocally influence each other in complex ways" (p. 92). Similarly, Heritage, Walqui, and Linquanti (2015) refer to them as language moments where teachers call students' attention to specific language features.

When micro-scaffolding, teachers draw from a wide range of internalized supportive behaviors to cue the learner at a moment's notice (Cole 2006). Micro-scaffolding relies on the hybrid language as visuals and tools are often added to natural language. Through a short exchange, the teacher links to the learner's background experience, cues the student to the new learning, and recaps the key point of the exchange at the end of the interaction (Hammond and Gibbons 2005). As this exchange occurs spontaneously and often tacitly (Polanyi 1967), there are many ways this can happen.

## The 5R Instructional Model

As researchers, we watched and analyzed hours of recorded lessons with EMLs, and we identified patterns that scaffolded the hybrid language within inquiry-based science instruction. As we tried to articulate the patterns, our ideas came together to form the 5R Instructional Model. Therefore, the model arose from theory and practice and had its foundation in scaffolding for EMLs.

The 5R Instructional Model is nonlinear and has five interwoven features (Figure 4.2). These features can be used to facilitate the inclusion and teaching of hybrid language within science content instruction. Four features (reveal, replace, reposition, and repeat) occur within the anchored context of the lesson. This anchored context always has the elements of a well-planned instructional event (macro-scaffolding) and the spontaneity of a lesson intervention (micro-scaffolding). Only the feature of reloading has a predetermined position in the lesson. Reloading is context reflective and used to revisit words and word meanings students have already encountered in previous lessons.

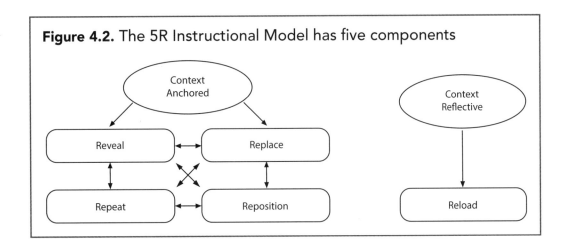

**Figure 4.2.** The 5R Instructional Model has five components

Chapters 5–9 describe the 5R Instructional Model in detail, but here we provide a brief synopsis to help you get a general idea of the components. These descriptions should give you a starting point from which you can think about the critical aspects of each feature, how the *R* is used, and why it is important for teachers of EMLs.

## Context Anchored

- **Replace** is used to provide EMLs with more precise ways of meaning-making across the hybrid language. Replacing builds on the background knowledge students bring to the lesson while developing additional ways to express academic meaning. Replacement mostly involves a one-to-one correspondence of a word, mathematical symbol, visual representation, or tool.

- **Reveal** is used to introduce the hybrid language when students have no informal language or conceptual understanding on which to build. This *R* is used often in science because processes, equipment, and concepts employed in the classroom do not have a generic or common name and are highly technical.

- **Repeat** is used to provide students with multiple opportunities to encounter and express meanings using all of the modes of the hybrid language. When engaging in repeated uses of the hybrid language, teachers and students use slight variations to communicate the same thematic relationships.

- **Reposition** is used to support EMLs as they move into the discourse of a community of practice and take the identity of a scientist. Repositioning offers students opportunities to talk like, act like, and look like scientists as they develop the meaning-making affordance of the hybrid language. It is the most complex component of the 5R Instructional Model.

## Context Reflective

- **Reload** is the purposeful revisiting of words previously encountered in science lessons. It explicitly supports EMLs in examining word relationships, word meanings, and word parts.

In the chapters that follow, we stress that the 5R Instructional Model is not confined to the teaching of vocabulary alone. Conceptual and multimodal language development go hand in hand—language development supports conceptual development, and conceptual development supports language development. Placing an emphasis on discrete pieces of language does not help students construct deeper meaning, whereas metalinguistic awareness helps provide a constellation of interrelated knowledge (Heritage et al. 2015).

As stressed by the *NGSS* and implemented in inquiry-based instruction, students interact with phenomena (manual-technical operations) to have first-hand experience on which they can build ideas. In addition, students use this experience to ask questions (natural language), carry out investigations (manual-technical operations), and collect data (natural language, mathematical expressions) that must be recorded (often in visual representations and natural language). Later, students analyze and interpret using mathematics (mathematical expressions) to communicate their findings (natural language, mathematical expressions, and visual representations). These multimodal representations are used with good guidance from you, the teacher.

As you read in more detail, you will notice several features about this instructional model. The first feature is that it is organic and cannot be thought of as linear. We will say this over and over because it is important that you do not think of this as a checklist but rather a new way to think about your planning. When and how the *R*s emerge in a lesson will depend on the situation, students, and event. The second feature is that, depending on the background knowledge of the students you teach, the same event may be a replace in one instance and a reveal, repeat, or reposition in another.

## Concluding Remarks

In this chapter, we outlined the 5R Instructional Model in general terms. Each remaining chapter begins with a scenario in which the *R* is evident in a science lesson. This scenario is followed by a rationale that expands the descriptions given here and provides supporting evidence for the *R*. In addition, each chapter further explores the *R* by giving examples of how it is manifested in each mode of the hybrid language of science. The chapter ends with a summary that highlights the major points.

# Chapter 5

# Replace

In his 11th-grade integrated physics and chemistry class, Mr. Jimenez teaches about models through a unit on alternative energy. Many of his students are newcomers and emergent multilingual learners (EMLs). Some of them are literate in their first language, many have emigrated from Central America, and some are refugees.

*Mr. Jimenez begins the lesson by asking students to help him decide the best design for a wind turbine to provide electricity for his house. After a discussion that results in students deciding that they need a model, he shows them a small, dynamic model of the turbine that will be used in the investigation. He suggests that students draw a picture of the turbine model and label it with names they think best identify each part. After students write their ideas, Mr. Jimenez draws a simple picture of a turbine on the whiteboard and asks students to call out the words they used to label its parts. He writes each student's suggestion beside the part being named. Students share a variety of ways to name the parts of the turbine, including petals, wingies, propellers, stem, tower, pole, energy producer, motor, and engine (Figure 5.1).*

*Mr. Jimenez tells his students that scientists have agreed to use the words* generator, blades, *and* tower *to refer to these components. Scientists use these names to make sure they are talking about the same thing. He writes each of these new terms*

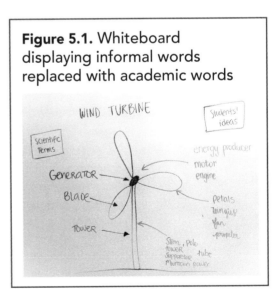

**Figure 5.1.** Whiteboard displaying informal words replaced with academic words

on the whiteboard beside the respective parts of the drawing and also adds them to the word wall.

Mr. Jimenez uses this activity to identify students' prior knowledge and to value the possible names they suggest for the components of the wind turbine. However, he also knows it is important for them to learn the appropriate scientific terminology. In addition, he is aware that language and cognitive development occur in parallel with language that supports students' learning of scientific concepts.

To have a record of the terms, students are encouraged to look at the model they will use and draw a more accurate illustration in their journals. When they complete the drawing, Mr. Jimenez asks them to label each part with its scientific name. Figure 5.2 is an example of a journal entry in which the student-generated informal terms are replaced with scientific terms. The students then design an investigation to answer the question of what is the best design to generate the most energy without damage to the wind turbine.

**Figure 5.2.** Student journal showing brainstorm of words that could be used to name the parts of the turbine (top picture) and replacement of informal terms with scientific terms after discussion (bottom picture)

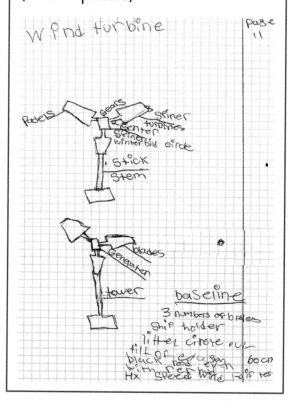

## Rationale

In this chapter, we discuss **replace** as one component of the 5R Instructional Model. For this *R*, you use students' previous understandings and informal language to scaffold relevant scientific concepts and hybrid language. Students are not asked to discard their informal language; rather, they are given language that situates them within a community of scientists. Replacement can occur within any of the four modes of the hybrid language, and, most often, replacement involves a one-to-one correspondence. In the previous scenario, Mr. Jimenez replaced the term *wingie* for the term *blade*. In addition, a replacement may occur across two modes, such as exchanging a symbol or an image for a word. For example, students learn water is also referred to as $H_2O$. This furthers the conceptual understanding that water is made of hydrogen and oxygen at a 2:1 ratio. Within mathematics, students learn to replace *is equal to or greater than* with ≥.

As students engage in first-hand experiences with natural phenomena (e.g., magnets, bird feathers, evaporation, dissolving), they spontaneously talk about what they observe. Informal discussions with unprompted talk often reflect the oral language (Gibbons 2015) discussed in Chapter 2 (p. 9). This type of language is important because "spoken language is the medium through which we reason to ourselves and talk our way through problems to answers. It is, for the most part, the medium in which we understand and comprehend" (Lemke 1989, p. 136). Lemke (1990) further asserts that students understand science best when they are allowed to first explain it using conversational language rather than complex jargon. He proposes that in the science classroom, students learn bilingually as teachers state and clearly signal that they are expressing scientific meanings using both informal and scientific language. Because members of the same social language communities tend to construct similar meanings from similar experiences, EMLs will ultimately develop the same canonical meanings that are recognized across the scientific community.

Students entering U.S. classrooms with little or no English do not arrive as a *tabula rasa* or blank slate. Instead, they enter school with knowledge and experiences that benefit them in learning science content and the hybrid language of its socially constructed community (Kutz 1997). When replacing informal language with the language of the science classroom, we connect the relevant background knowledge and experiences EMLs bring with them as we bridge new conceptual understandings and features of the hybrid language. In addition, we are not asking EMLs to undervalue conversational language or privilege science language. Instead, through replacement activities, we aim to support learners in developing a larger repertoire of ways to express meanings that are more appropriate to the science contexts in which they are used.

Although students initially understand conceptual and linguistic relationships using more spoken-like speech, they "substitute scientific, technical terms for colloquial words" (Lemke 1990, p. 173) as they move into academic language. It is important to note that meanings expressed in spoken-like language can obscure underlying differences that are important in science discourse (Gee 2004). Furthermore, misconceptions can be brought to or developed within the classroom through informal language. In the scenario at the beginning of this chapter, students suggest the words *motor* and *engine* when first naming the parts of the wind turbine. Though these words conceptually relate to the notion of a generator, they do not take into account the essential characteristics of the generator as a device that converts mechanical energy into electrical energy. By contrast, a motor converts electrical energy into mechanical energy, and an engine converts heat energy (steam) into mechanical energy.

As teachers help EMLs explore the function of generators within the context of a unit on wind turbines, students are also developing the conceptual understandings that allow them to recognize important characteristics to distinguish a generator from a motor. Gee (2002) also reminds us that everyday language creates important patterns and generalizations. However, scientific discourse serves to create knowledge claims by establishing differences and using oral or written language to create logical relationships (e.g., cause and effect, contrast, conditions, taxonomy).

# Replacing the Hybrid Language of Science
## *Natural Language*

In the opening scenario, we observed Mr. Jimenez as he helped EMLs replace informal concepts and words with new scientific understandings and academic language. Scientists use unique names to avoid ambiguity when communicating. Biologists, for example, need to distinguish among closely related organisms and avoid common names (e.g., cat) because they do not convey the specific meanings of a particular species (e.g., *felinae*, *pantherinae*). To precisely convey such meanings, the science community has a long history of developing specific taxonomies for naming, classifying, and describing specimens based on commonalities.

It might not be apparent to a novice observer, but Mr. Jimenez entered this lesson with a plan. As he prepared his lesson, he designed an interaction that would allow him to use what EMLs knew about the parts of a wind turbine. He then replaced informal language by building on previous knowledge and language. Furthermore, Mr. Jimenez made a point of supporting students as they considered how scientists use words that are different from those used in informal language to describe wind turbine components. This is an example of planned macro-scaffolding critical to the 5R Instructional Model.

Another way teachers can replace language for EMLs is by tapping their expertise as translators and interpreters. EMLs often think that science is hard because they do not yet have a full grasp of English. They do not understand that even for their fluent English counterparts, the language of science is also a new language. Documenting the skills that EMLs develop as they translate and paraphrase English for their families on a daily basis, researchers emphasize the benefits of helping students recognize and value their skills as translators within school settings (Martínez et al. 2008; Orellana and Reynolds 2008).

In another wing of the school, we observed Ms. Sabala work with EMLs on a lesson that tapped students' translating skills as they replaced words across informal and academic language registers. To guide the activity, Ms. Sabala posted a series of directions on the interactive whiteboard (Table 5.1).

After this activity, Ms. Sabala and her students discussed some of the challenges they encountered when using this translating strategy. Ms. Sabala asked students to consider which process was more challenging—translating from informal to academic or from academic to informal—and to provide a rationale for their answers. Figure 5.3 is an example of a journal entry completed by one of her students as part of this metalinguistic process.

Even though we focus on the teacher's role in scaffolding opportunities for students to replace natural language, developing metalinguistic awareness helps students independently make replacements in their own speaking and writing. For example, when writing a procedural account of a Kastle-Meyer test (a presumptive test giving a pink color with blood), a student in Ms. Sabala's classroom self-corrected her journal entry (Figure 5.4, p. 50) by replacing the word *water* with *phenolphthalein*.

In class, students used phenolphthalein to determine whether the samples they had collected at a crime scene were blood or not. They had been directed to use water on a cotton swab to lift

**Table 5.1.** Sort and Match Informal and Academic Language Directions

**Sort and Match**

Take one word out of the envelope and say:

My word is …

I think my word means …

I think my word is academic or informal because …

For every academic word, try to find an informal word and write it on the opposite side of the index card.

   Example:

  *observe* (academic)

  *look at* (informal)

For every informal word, try to find an academic word and write it on the opposite site of the index card.

   Example:

  *change* (informal)

  *transform* (academic)

As a group, discuss whether you agree or disagree.

**Figure 5.3.** Example of a student's journal reflection as part of a metalinguistic assignment

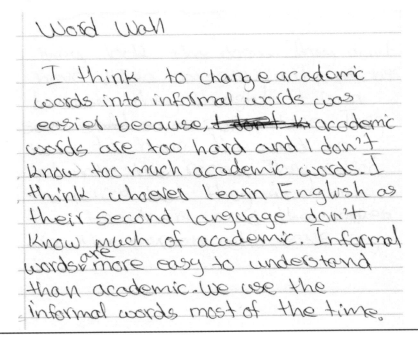

> Word Wall
>
> I think to change academic words into informal words was easier because, ~~I don't k~~ academic words are too hard and I don't know too much academic words. I think whoever learn English as their second language don't know much of academic. Informal words are more easy to understand than academic. We use the informal words most of the time.

**Figure 5.4.** Journal example and transcript of a written replacement—*phenolphthalein* for *water*

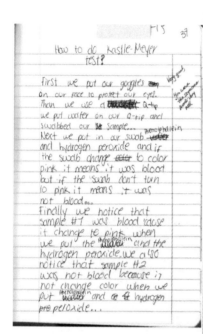

How to do a Kastle-Meyer test?

First we put our goggles on our face to protect our eyes. Then we use a Q-tip We put water on our Q-tip and swabbed our sample... Next we put in our swab ~~water~~ phenolphalein and hydrogen peroxide and if the swab change to color pink it means it was blood but if the swab don't turn to pink it means it was not blood... Finally we notice that sample #7 was blood 'cause it change to pink when we put the ~~water~~ phenolphalein and the hydrogen peroxide. We also notice that sample #2 was not blood because it not change color when we put ~~water~~ phenolphalein and hydrogen peroxide.

and moisten a dry sample and then drop phenolphthalein onto it to watch for a reaction. In the journal entry, the replacement is significant because the use of phenolphthalein also reflects the writer's developing conceptual understanding of the technical term. Water and phenolphthalein share similar characteristics—they are both clear, odorless liquids—and in her journal entry, the student used the term for the clear liquid she was most familiar with. However, with self-editing, she realized the difference and replaced the incorrect use of *water* with the term for the reactive liquid—*phenolphthalein*. Had the student not replaced *water* with the more technical word, then her understanding would have been construed as a misconception.

## Mathematical Expressions

In Chapter 2, we discussed mathematical expressions as being typological and topological. One form of replacement is to help students move from typological expression (usually stated through natural language) to topological (usually expressed through numbers and symbols). For example, typological to topological replacement occurs when the teacher supports EMLs in replacing words such as *cold* and *hot* with precise temperature measurements.

When visiting Ms. Kris's mathematics classroom, we observed a replacement lesson as she was introducing early algebraic thinking to her EMLs. Ms. Kris helped students replace words such as *sum* with the mathematical symbol for plus (+) and *niece/nephew* with acceptable mathematical shorthand (*n/p*). Although you may most often use *x* and *y*, any symbol will work.

To start the lesson, Ms. Kris used a problem from the dedication page of *Math Curse* (Scieszka and Smith 1995, p. 1), one of her favorite read-aloud books:

> If the sum of my nieces and nephews equals 15, and their product equals 54, and I have more nephews than nieces, how many nephews and how many nieces is this book dedicated to?

Ms. Kris first guided students into replacing natural language with an equivalent mathematical expression that would allow them to solve the problem (Figure 5.5).

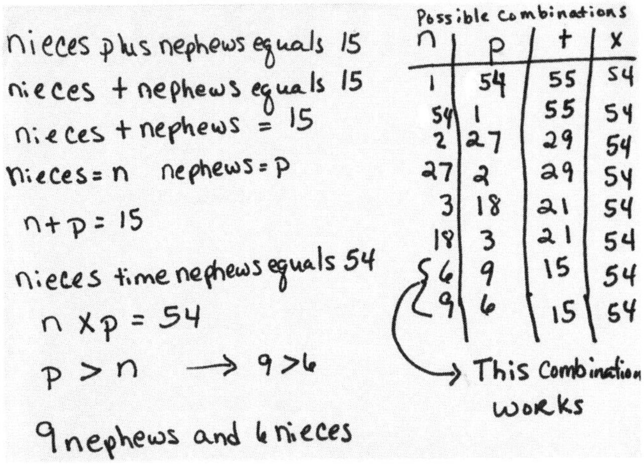

K: What does the word *sum* mean?

S1: It means to add.

K: So how would I write "the sum of my nieces and nephews equals 15"?

S1: Nieces plus nephews equals 15.

[Ms. Kris writes, "nieces plus nephews equals 15."]

K: I can replace the word *plus* with the plus sign (+).

[Ms. Kris then writes, "nieces + nephews equals 15."]

At this point in the lesson, Ms. Kris used the exact words from the book to get students thinking about the mathematics problems and how to replace natural language with mathematical expressions. To continue probing students, Ms. Kris selected to replace another part of the sentence.

**Figure 5.5.** Example of students replacing words with symbols based on *Math Curse* (Scieszka and Smith 1995)

K: Is there another word in this sentence we can replace with a mathematical symbol?

S2: I think you can change *equals* to =.

Ms. Kris knew students most often experience mathematical problems as natural language and then translate the problem into mathematical expressions to find the solution. By having students talk about what can be replaced, she scaffolded their ability to move from one mode of the hybrid language to another. Ms. Kris continued the lesson.

K: Is there another way to write the sentence other than writing "nieces and nephews" out each time? [There is no response from students, so Ms. Kris continues probing.] Mathematically, what are nieces and nephews?

S2: Variables?

[The whole class agrees.]

S3: I know we can change nieces to *n*.

K: OK, then what about nephews?

S2: We can't make it an *n* again.

S4: Could we make it a *p*?

K: Does it matter what letter we use to replace *nephews*?

S1: No, it doesn't matter.

Ms. Kris continued working with students to change the word problem into a symbolic form that is used in mathematics.

K: Look at the second part of the sentence. It says, "and their product equals 54." What does the word *product* tell us to do in this sentence?

S5: It means to multiply.

K: So how would I write that using a mathematical expression?

S5: You write *n* times *p* equals 54.

K: Work with a partner and write this sentence using all mathematical symbols.

After a pause, Ms. Kris borrowed a student paper and wrote the following on the whiteboard:

Nieces plus nephews equals 15          Nieces times nephews equals 54

$n + p = 15$                           $n \times p = 54$

Ms. Kris built on students' prior knowledge as she helped them determine the appropriate mathematical symbols for key words in the written texts. First, students had to determine what *sum* and *product* meant. Then they replaced the mathematical symbol in the first statement (nieces + nephews = 15). Next, students replaced the words *nieces* and *nephews* with letters ($n$ = nieces and $p$ = nephews). Finally, they produced equations to represent the words within the story ($n + p = 15$ and $n \times p = 54$).

We selected this example because it illustrates replacement while also demonstrating how to help students develop metacognitive and metalinguistic skills. Ms. Kris was able to demystify the process of going from a written problem (natural language) to a mathematical expression by having students think and talk about how terms are replaced with symbols. As we discussed in Chapter 4 (p. 35), this naming of a thing or process provides students with the ability to recognize and use the symbol or process in other contexts.

Another example of replacement is seen when Ms. Kris recognized that EMLs had misconceptions concerning fractions. The habitual misuse (in natural language) of terms that have specific conceptual meanings may have led to mathematical misconceptions. This is often observed in how teachers and students use the terms *fraction, numerator,* and *denominator.* Most children, and even adults, define a fraction as a *part of something* as opposed to the mathematical understanding that a fraction is an *equal part of a whole.*

Ms. Kris spent time engaging students in hands-on activities using manipulatives (pattern blocks, double-sided counters) to determine an understanding that *something* can be replaced with the concept of a *whole.* She then replaced students' misunderstanding that a denominator is "the number on the bottom" with the conceptual understanding that a denominator is the number that represents the equal number of parts the *whole* is broken into. Once students recognized that *denominator* has a precise meaning, they began to think about what the numerator represents.

Ms. Kris continued to help students move from thinking of the numerator as the "number on the top" to recognizing it as the number of equal parts in which they are interested. Neither number means anything without the other because a fraction is a relationship of part to whole. These definitions continue to build on students' common, informal language, but they replace previous misunderstandings with more precise conceptual understandings and language.

## Visual Representations

As with natural language and mathematical expressions, context helps dictate the type of visual representations used to communicate meaning. Within the science classroom, simple sketches are often used to capture initial impressions, but these may be replaced with more complex representations. In the opening scenario, Mr. Jimenez asked students to draw a rough sketch of a wind turbine in their journals, thus giving them a visual reminder of the structure of the turbine. After discussion and the replacement of technical terms, he asked them to replace the original sketch with a drawing that more closely represented the model being used. This replacement resulted in a visual representation that incorporated natural language labels and provided students with a resource for future use.

Another example of replacement within visual representations occurred as part of the CSI unit in Ms. Soto's classroom (Chapter 2). In this case, her objective was for EMLs to understand the need to have an accurate depiction of a crime scene. To scaffold this notion, the first activity she planned was to have students visit the crime scene. Students were asked to sketch the scene to capture important details. Figure 5.6 is an example of a graphic representation created by one of the students in her class.

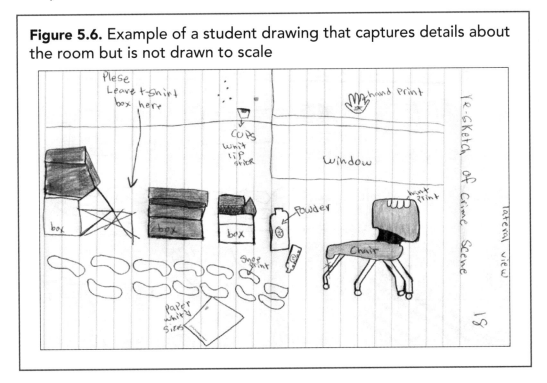

**Figure 5.6.** Example of a student drawing that captures details about the room but is not drawn to scale

Ms. Soto then guided students to recognize variations among students' representations. She had students return to the scene with measuring tapes and collect data they thought would be important to know about the size and shape of items in the room to produce an accurate representation of the crime scene.

When students returned to the classroom, they brainstormed ways to represent the large crime scene on small pieces of paper. They had room measurements and sketches of how items were placed, so they replaced the freehand drawing of the crime scene with a scaled drawing. This is an important replacement because the notes taken by scientists are often recorded in scaled dimensions. From the replacement, they began to understand that the scaled representation is an example of the discourse used by scientists.

These examples show how you can support replacement in students' visual representations by helping them move from informal/less accurate depictions to more accurate representations, as appropriate. As you think about replacement, you can help your students understand (1) which representation is most appropriate and (2) how to make a more precise depiction.

## Manual-Technical Operations

Replacements with manual-technical operations involve exchanging less sophisticated tools that use gross hand movements with more advanced tools used by scientists in field and laboratory settings. Having better or more sophisticated tools aids the meaning-making process because, like mathematical expressions, they allow for more precise communication. The ability to use the correct tool in the right context at the right time is part of the social language of science that separates the expert from the novice.

An example of replacement occurred in Ms. Tatum's class when she used scientific measuring tools. Ms. Tatum knew that her students had used a double-pan balance in elementary school to see the relationship between two different weights. In mathematics, this manual-technical operation is used to teach the concepts of the commutative property and equality. Ms. Tatum wanted students to develop an abstract concept of weighing and be able to collect more precise information. Therefore, she replaced the double-pan with a triple-beam balance. This manual-technical operation required students to move the weights to a position of balance and determine the reading. She knew that in high school this tool would be replaced once again because students would need greater precision. If students did not have background instruction in measuring weights, then a digital scale might seem like magic.

Another replacement of a tool was seen when Ms. Soto helped students move from using a pencil and graph paper to a graphing calculator (Figure 5.7). A graphing calculator allows students to input data manually or through link ports and then manipulate the data to show different relationships. The use of calculators allows students to make meaning of data quickly. It also provides students with the ability to manipulate data to communicate new relationships.

Replacing requires students to think about a more appropriate tool to use under different situations. Inquiry-based, manipulative-rich lessons—whether in science or another subject—provide students with opportunities to use tools to make meaning and communicate with classmates and you.

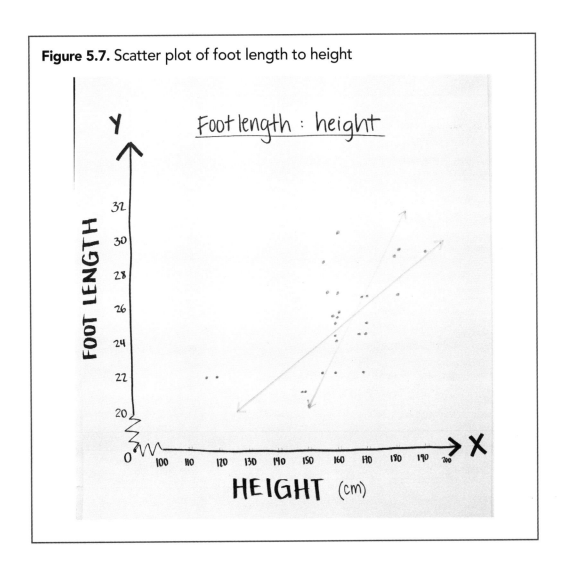

**Figure 5.7.** Scatter plot of foot length to height

## Concluding Remarks

Replacing serves to facilitate the development of more precise ways of meaning-making using the hybrid language of science. Replacements mostly involve a one-to-one correspondence of a word, mathematical symbol, visual representation, or manual-technical operation. They occur as teachers scaffold the development of the hybrid language through planned instructional activities or take advantage of spontaneous learning opportunities within a lesson.

# Chapter 6

# Reveal

**M**s. Johnson, an eight-year teaching veteran, knows that her emergent multilingual learners (EMLs) bring knowledge to class. In the following scenario, she teaches the properties of light and the practice of carrying out an investigation. She also taps into prior knowledge as she introduces them to a new lab tool.

*Students in a physics class have just finished using a prism to separate light generated by a regular light bulb. Ms. Johnson wants EMLs to experience this same phenomenon using complex equipment. She moves to the back of the room where several pieces of equipment are positioned on laboratory tables and picks up one of the tools. She shows it to the class and asks, "Have you used one of these? Do you know the name of this piece of equipment?" There is a general murmur toward the negative, so she continues.*

*Pointing to the tool, she says, "This is a spectroscope. Watch as I use the spectroscope. I want you to notice what I do so you will be able to use it quickly for your investigation. I also have written directions, in case you need them." Because the tool is expensive and has a specific way to be operated, Ms. Johnson demonstrates the correct way to use the spectroscope.*

*At the end of the demonstration, she again provides the name of the new tool by asking, "What is this called?" and responding, "Yes, this is a spectroscope." She then places a paper strip with the new word on the word wall. "Let's think about its function in relationship to parts of this word we might have seen before. What do we know about the word* scope*?"*

*One student says, "A scope can be put on a rifle."*

*Another responds, "Not in school. In biology, we have microscopes!"*

*Ms. Johnson then asks, "What is common between a gun scope and a microscope?"*

*Several students respond at once: "They help you see." "They make things bigger." "They bring what you want to see closer to you."*

*Ms. Johnson guides students in a discussion about the word* scope *(meaning "the range of view") and how its meaning relates to the topic of light and optics. She explains that the spectroscope, like the prism, breaks white light into a range of the visible spectrum—hence,* spectro—*allowing them to see the color bands. Ms. Johnson asks students to get in their investigation groups and practice using the spectroscope. After all students have an opportunity to use the tool, she proceeds to discuss the follow-up activity.*

## Rationale

In this chapter, we introduce **reveal** as one component of the 5R Instructional Model. A critical aspect of this *R* is that it supports EMLs as they develop language for new concepts within the context of the science lesson. When revealing any mode of the hybrid language of science, you provide students with experiences to build new science understandings and develop the language they need to make sense of these experiences. Ms. Johnson, in our opening scenario, builds on students' understanding of light refraction as she explicitly introduces new language and lab equipment.

Underpinning reveal as one of the components of the 5R Instructional Model is an idea introduced in Chapter 3 (p. 21), when we discussed that optimal teaching occurs in Cummins's (1981) Quadrant B. When Ms. Johnson targets instruction to occur in Quadrant B, she engages students in cognitively demanding tasks yet scaffolds language to make these tasks comprehensible. Entering the lesson, students have not had many experiences with the ideas Ms. Johnson puts forward in the lesson. Understanding light refraction is cognitively demanding. Ms. Johnson's demonstration of a spectroscope, along with her teacher talk, provides a rich context for students to comprehend the challenging concepts introduced in the lesson. During the lesson, she pays attention to the natural language and manual-technical modes of communication.

In many science classrooms, the presentation of new language comes in the form of front-loading vocabulary activities. When front-loading vocabulary, teachers select and preteach key terms and definitions that students will encounter. Teachers often engage in frontloading activities with the well-intentioned belief that this practice provides EMLs with the background knowledge and contextual support they need to fully engage in the science lesson. Front-loading vocabulary practices, however, often morphs into the preteaching of content (Livers and Bay-Williams 2014). Specifically addressing the practice of front-loading vocabulary in mathematics lessons, Livers and Bay-Williams further argue that this practice not only robs students of opportunities to explore concepts in concrete and meaningful ways but also takes time away from exploring mathematics tasks.

In the 5R Instructional Model, revealing is different from front-loading. We conceptualize revealing as occurring within the lesson as students construct new science understandings and develop the hybrid language of science. In Chapter 2 (p. 9) and Chapter 3, we considered how science learning is related to the learner's ability to make meaning and use the social language of the science community. We further expanded this idea as we discussed how scientific concepts emerge as learners interact with natural phenomena and use natural language, mathematical expressions, visual representations, and manual-technical operations to construct meaning within the science community.

The reveal feature can occur at any point within a lesson. Although it is most often planned (macro-scaffolding), it can occur spontaneously (micro-scaffolding). At this point, we want to reiterate that the 5R Instructional Model is not linear. Replace (introduced in Chapter 5, p. 45) and repeat (discussed in Chapter 7, p. 67) do not come into a lesson at the beginning with reveal and reposition (discussed in Chapter 8, p. 75), which comes in the middle or end of a lesson. In fact, the five features may flow into one another seamlessly. The *R*s instead serve as a deliberate framework to scaffold the development of multimodal language within the science classroom.

## Revealing the Hybrid Language of Science
### Natural Language

In the opening scenario, Ms. Johnson explicitly revealed a new science word—*spectroscope*—while she also revealed how to use this new manual-technical tool to construct understandings regarding the properties of light. Similarly, in your classroom, students constantly encounter new ideas for which they have no words. In fact, in an analysis of secondary science textbooks, Yager et al. (2009) determined that there are more new terms in a first-year biology course than in a first-year foreign language course. Given the prevalence and significance of new terms in science, we focus on ways teachers reveal new words and word meanings as they scaffold new understandings in the classroom.

When working with classroom teachers, discussions about word learning are often framed around the teaching of vocabulary. These vocabulary teaching practices, as we previously discussed, are too often reduced to the front-loading of new words and their definitions. Although definitions can be useful, they are not enough. Researchers consider word knowledge to be a continuum that ranges from not knowing a word, to having a general sense of a word's meaning, to recognizing a word in certain contexts, to the development of conceptual knowledge that allows learners to make connections to other words (Blachowicz and Fisher 2006; Graves, August, and Mancilla-Martinez 2013; Haug and Ødegaard 2014). Researchers also emphasize that understanding of a word's meaning, rather than occurring as a one-shot deal, develops incrementally over time as learners have multiple opportunities to encounter that word and its definitions in different contexts (Scott 2005; Stahl and Nagy 2006). These notions regarding the development of words, word knowledge, and word meanings will be further examined when we consider three additional *R*s within the 5R Instructional Model: repeat, reposition, and reload.

Germane to this discussion is an understanding of the relationship among words, word meaning, and concept development. Because concepts do not have a one-to-one relationship

with individual words, in the science classroom, meaning is best expressed in terms of thematic patterns. Lemke (1990) explains, "To get the meaning of the whole, you need to know more than the meaning of each word: you need to know the relations of meaning between different words" (p. 12).

Gibbons (2015) views the relationship between content knowledge and new language development as reciprocal. In the science classroom, "content learning provides a 'hook' on which to hang language development, and vice-versa" (p. 209). Rather than front-loading isolated terms and their definitions, instruction focuses on helping EMLs learn new words and meanings as they develop new conceptual relationships about the phenomena being studied.

In Ms. Johnson's class, a new word—*spectroscope*—is taught within the context of other related words (e.g., *refraction, wavelength, spectrum*). Similar to Cummins's (1981) notion that optimal teaching occurs in Quadrant B, Gibbons (2015) notes that effective teachers do not shy away from introducing complex ideas and concepts. Instead, they support EMLs as they engage in challenging concepts through demonstrations, the presentation of a variety of materials, and the use of natural language to combine words with other meaning-making modes.

In addition to introducing new words, Ms. Johnson engages students in developing metalinguistic awareness (i.e., talk about language, not just language use). In the scenario, we observe how, after introducing the use of a spectroscope, Ms. Johnson explicitly returns to the word *spectroscope* and asks students to notice word parts as she guides them through a discussion regarding the function of *scope* within the new word. When teachers promote word consciousness, they help EMLs become aware of words, develop an interest in learning words and their meanings, and develop knowledge about words. This type of word consciousness is viewed as critical for vocabulary development (Graves et al. 2013; Scott and Nagy 2009; Stahl and Nagy 2006). Teachers also support metalinguistic awareness by helping students understand that the words they use in the science classroom reflect the particular social language of the discipline. As such, scientists use words that are not used in other situations, so students must learn these words if they are to talk like scientists.

Ms. Johnson carefully plans (macro-scaffolding) which words she will reveal to students in a lesson. As she does this, she also identifies concepts she anticipates will be new to students and are important for them to understand so they will be able to engage in a unit of study. This initial process allows her to consider how she can deliberately scaffold these new words within the context of lessons she plans to teach. She also writes the new words on paper strips that she adds to the classroom word wall after they are introduced. These serve as visuals she can use and refer to throughout the lesson. As a result, words become tangible objects that can be manipulated when engaged in reloading activities (discussed in Chapter 9, p. 89).

Though the context of a lesson affords EMLs with an opportunity to develop new words and word meanings, research indicates that more words occur in written form than in oral form (Freeman and Freeman 2009; Stahl and Nagy 2006). Consequently, teachers promote the development of academic words and word meanings through reading. Anderson and Nagy (1993) report that students can learn an average of 2,000–3,000 words per year by reading.

Similarly, Krashen (2003) concludes that, like native speakers, EMLs make vocabulary gains in second language acquisition by reading.

One way of integrating reading to support word learning in the science classroom is through trade books (Fang 2013; Fang, Lamme, and Pringle 2010). Saul and Dieckman (2005) also call our attention to the value of trade books to introduce children to specialized terminology they are not likely to encounter in their everyday experiences with the natural world. These specialized words are embedded in contexts that are rich in descriptions, examples, and illustrations. To promote the use of trade books in science classrooms, the National Science Teachers Association publishes an annual list of outstanding science trade books for use in K–12 settings *(www.nsta.org/publications/ostb)*.

For trade books to be used effectively, they must serve as tools to support inquiry in the science classroom—not merely as methods to gain factual information. Researchers raise concerns about the misalignment between many trade books and the principles underlying current understandings of the nature of science (Ford 2004, 2006; Zarnowski and Turkel 2013). Ford's (2004) content analysis of science trade books indicates that, even though trade books highlight fun experiments, the hard work of scientists, and an appreciation of nature, they still fall short in how they describe the development of knowledge, the way the scientific community works, and how readers should ask scientific questions.

Despite these shortcomings, trade books can serve your students as tools for developing, researching, and investigating questions that are not easily tested within the contexts of empirical experimentation. Classroom teachers can encourage the use of trade books to support students as they learn to critically evaluate text information, compare text resources, and build explanations using multiple sources of evidence. Ford (2006) contends that the instructional context in which trade books are used will ultimately determine their utility in the inquiry classroom. If positioned as one of many tools available in constructing scientific understanding, then trade books indeed have a place in inquiry-based science.

Teachers often integrate trade books through the use of book clubs organized around text sets. A text set is a collection of conceptually related trade books connected to the inquiry topic being investigated in class (Fries-Gaither and Shiverdecker 2012; Short, Harste, and Burke 1996). There were six to eight copies of each book available, so students gathered in small groups to read and discuss the books they selected. Text sets, because they are intertextually linked around a similar topic, offer EMLs the opportunity to see new science words in relation to other words within the same thematic patterns. The use of trade books, however, is not limited to science. Teachers often make use of mathematics trade books to support EMLs in developing ways of expressing meaning using mathematical expressions and visual representations.

## Mathematical Expressions

Lemke (1990, 2002) reminds us that mathematical symbols and expressions allow for a precision that is impossible with other modes of the hybrid language. In the scenario that opened this chapter, Ms. Johnson revealed the spectroscope to students, who then gained experience with the tool (manual-technical operations) while making meaning about the splitting of white light into the continuous spectrum. However, as beneficial as this is, it did not reveal every

wavelength. Through an activity looking at hydrogen gas, she revealed the Balmer formula:

$$\lambda = \frac{91.1nm}{\left(\frac{1}{m^2}-\frac{1}{n^2}\right)},$$

where $m$ = 1, 2, 3 … and
$n$ = any integer > $m$

This formula allowed students to describe every wavelength of hydrogen's emission spectrum, thus helping them develop an understanding of energy and wavelengths. Once the formula had been revealed to students, they could use it in follow-up lessons.

A second example of revealing mathematical expressions takes us back to Chapter 2, where Ms. Soto's biology students were engaged in a CSI unit. As part of that unit, students found footprints left at the crime scene and needed to

**Figure 6.1.** Students examining the relationship between foot length and height

**Figure 6.2.** Chart with various ways of representing mathematical expressions depicting the relationship between foot length and height

foot length
height

foot length  to  height

foot length : height

$\frac{25}{163}$ ≈ 153 ≈ 15.3%
                          15%

25 to 163  | 153 : 1000
25 : 163   | proportional

know more about how they could be relevant to the identification of a suspect. To learn more, they made prints of their own shoes by stepping on ink paper. These prints were then used to help students develop an understanding of how a foot length could help identify possible suspects. Ms. Soto asked students to use those prints and line up according to the length of their feet (Figure 6.1). Students then positioned themselves in a line from shortest to longest foot length and noticed that there was a relationship between foot length and height. In general, as students' feet got longer, their bodies got taller.

Ms. Soto revealed that in mathematics there is a way to express what they observed in the human foot line. As seen in Figure 6.2, she used the students' information to show the relationship as foot length (F) to height (H) (foot length : height, F:H). This hands-on experience revealed the formula to express the relationship that investigators use to determine the height of a suspect when given a footprint.

Following the lesson where Ms. Soto revealed the formula, she had students complete a "What I Did"/"What I Learned" reflection (Figure 6.3 and Table 6.1). Here we note how, on the "What I learned" side of the entry, the student demonstrates her new understanding of the relationship between foot length and height by using a more precise mathematical formula.

Another example of reveal can be seen as students move into higher-level science and mathematics. They learn to express scientific notation, which is a short-hand expression used when studying phenomena that are very small or very large. A number can be expressed using multiple digits (e.g., 0.0000000000007 or 7,000,000,000,000). The science teacher must reveal scientific notation as a way to provide information in a truncated manner (e.g., $7 \times 10^{-13}$ or $7 \times 10^{12}$).

**Table 6.1.** Transcription of Journal Entry

| What I Did? | What I Learned? |
|---|---|
| We watch a video on how to take photographs of a footwear print. | I learnt the formula of slope = |
| We lined up our footwear print from smallest to biggest. | $$slope = m = \frac{y_2 - y}{x_2 - x_1} =$$ change in footlength / change in height |
| We made a scatterplot of our heights and footlength, then we find the mean. | I also learned how to say and read decimal numbers. I ~~found out~~ measured my height and foot length. |
| We had a positive trend of our scatter plot. | |
| We also did a graph to draw a sketch of our measured objects. | I learnt that the line of bestfit is a line that fits between points on a scatterplot. |

**Figure 6.3.** Student journal page identifying mathematical expressions revealed in prior lesson

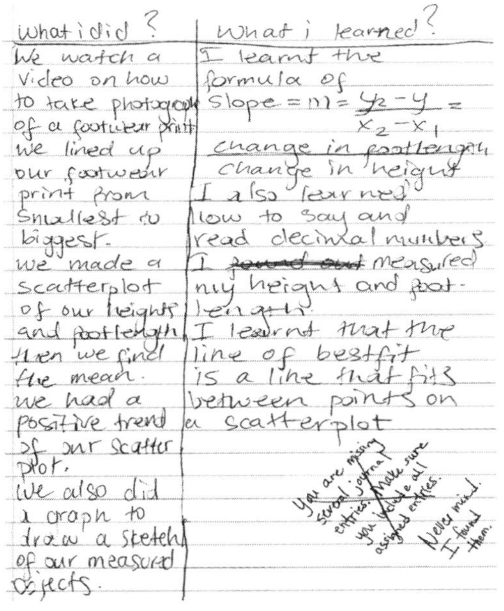

## Visual Representations

Visual representations provide information that cannot be conveyed by other modes of the hybrid language. Textbooks are filled with visuals because the authors and publishers are aware that natural language is not always sufficient for revealing new concepts. For example, a photograph is used in place of a phenomenon when obtaining the real thing is not possible or would be too small to see or when a description would be insufficient. Therefore, we use photographs to reveal structures such as the surface of Mars or microscopic viruses.

There are times when photographs are not the most appropriate visual because they do not allow a novice to distinguish specific features. This is particularly true for something that is unfamiliar, such as cell structures or the surfaces of the other planets. By contrast, an artist's representation (drawing) often pinpoints relevant parts of interest by highlighting or magnifying them. In a life science unit, a photograph of a typical cell has dark and light areas

that can be difficult to distinguish even with labels. This lack of clarity is partially because students do not have background knowledge of the organelles within a cell. A stylized representation, however, can show the organelles in ways that more easily reveal the structures for students.

We also reveal how authors use text features through read-alouds and when working in small reading groups. Figure 6.4 displays a student journal entry after a lesson in which the teacher and students discussed visual text features found in a science textbook. In this journal entry, the student expresses what he learned about how authors use visual representations to help readers navigate a text. To support EMLs in taking full advantage of additional meaning-making cues that visual representations provide, it is important to reveal why certain forms are better at conveying meaning than others.

Another visual that needs to be revealed are data tables. To support students' understanding of how to read and write data tables, teachers must explain how the data are presented. This includes an examination of how items within similar categories are entered in columns and rows. As students work with their own data tables (see Figure 3.3, p. 33) or as they encounter new tables in trade books or textbooks, it's important to work on reading strategies that help them locate, compare, and contrast different data points.

**Figure 6.4.** Student journal entry demonstrating a new understanding of the role of text features

> Day 3      Page 5
>
> features of ~~text~~
>
> Today we learned about features of text with Mrs. ____ and I am going to talk about how features of text help me ~~a~~ write better and understand what we are reading.
>
> There are a lot of things about features of text there has to be an index, a table of content/contents glossary, cover illustration, title and subtitle, illustration, photos, color full text, bold text, italics, picture caption, paragraph, graphs, charts, and maps, page number, back cover information and a key word box.
>
> Those are the things ~~to~~ that you might see ~~m~~ when you are ~~wrt~~ reading a book and when you ~~right a~~ write a book. Thank you bye.
>
> Which features of text will you use in your article?

## Manual-Technical Operations

As indicated in previous chapters, meaning-making in science classrooms involves using specific equipment or tools in an appropriate way and at the appropriate time. Although we strongly suggest that students be allowed to muck around with equipment and procedures to figure out how they work, we know there are some pieces of equipment that are much too expensive or dangerous to let students simply "play" with. There are also precise procedures that have been worked out over many years of experimentation that would take too much time for students to figure out on their own. This was the case with the spectroscope noted in the opening scenario. For this reason, you may have to reveal these manual-technical operations to your students.

The following transcript of a lesson provides an example of a classroom teacher revealing manual-technical operations as she introduces a voltmeter for students who are learning about electricity. The teacher not only reveals the name but also reveals how to use the voltmeter.

T: A volt is the unit—like inch, centimeter, pound, or gram—that is used to measure the flow of electricity. By using a voltmeter, we can get a number that is accurate for telling us how much electricity is flowing through the wire. [The teacher puts the voltmeter on the document camera so it is projected at the front of the room.] Lisbet, please come point out interesting things you see on the voltmeter.

S: Here is a dial. It has different settings.

T: What else?

The teacher collaborates with Lisbet to help her reveal the various parts of the tool and how they work. Together, they reveal the procedure for using a voltmeter by demonstrating how to connect it to the electrical source, how to set the dial, and how to read energy output due to the manipulation of different variables. Students then use the voltmeter correctly to enhance their understanding of change in energy output. The essential aspect of this lesson is that the piece of equipment is introduced and used within the context of the lesson. Students then conduct several investigations requiring the correct use of the voltmeter.

If your objective is to separate component substances from a liquid mixture, you may want students to discuss all of the ways they think a mixture could have its parts separated. After several ideas have been put forward, you can build on these and further explain the process of distillation. Because the tubing and glassware found in a standard distillation apparatus is complicated and needs to be assembled in a particular way, you should show students how to assemble the apparatus. This reveal introduces students to the situated context necessary for use of the equipment, whereas using the equipment provides students with meaning-making functions.

An additional example comes from Ms. Soto's classroom (Chapter 2), where EMLs studied genetics and heredity. When students were studying DNA and engaged in preparing an electrophoresis gel, Ms. Soto first replaced the word *dropper* with *pipette* as students examined the tools they would use to separate the DNA fragments. She recognized that students who were familiar with droppers and could use them correctly may not have known how to use the more delicate tool. Thus, a reveal was needed.

Ms. Soto helped students with the positioning of their fingers on the pipette and demonstrated how to get the suction correct. Precise positioning, followed by an exact movement upward, created a suction that moved a DNA sample into the pipette. Depositing the sample into a well located at the end of the gel required reversing the first movement. Once these movements were modeled for students, they had to practice several more times before they could easily re-create the procedure.

## Concluding Remarks

Reveal is used to introduce EMLs to new concepts using the four modes of the hybrid language. The more knowledgeable other introduces, within the context of the lesson, the concept for which the student has no prior knowledge on which to build. Unlike front-loading, revealing occurs at the point in the lesson when the knowledge is needed to move forward. This *R* is used quite often in science because some processes, equipment, and concepts do not have generic or common names already familiar to the students.

# Chapter 7

# Repeat

Mrs. Malo, who we introduced on page 32, works in a district with a high population of emergent multilingual learners (EMLs). She teaches erosion as part of a unit on Earth's systems and the practices of carrying out an investigation. In the following lesson, she provides students with recurring opportunities to engage with science concepts and the hybrid language required to comprehend and express their scientific understandings.

*Recognizing that inquiry involves asking and designing ways to answer questions, Mrs. Malo introduces an erosion unit by posing the following question: "What will happen to my yard, a sandy hill, if the meteorologist is correct and we have a gentle rain?" As EMLs think about this question, Mrs. Malo asks them to design a way to find an answer. She and her students decide that a model of the yard can be constructed using a stream table and sand. The students determine how to set up their model's baseline—how much sand, the angle of the slope, the position of the rain cloud—so each table has the same model. In addition, each investigation is compared against the baseline and one another.*

*To test their ideas, the class revisits the concepts of a fair test, a baseline design, and variables as they conduct three trials in which only one variable (rainfall) is tested. Each repeated trial requires students to set up the model using predetermined dimensions. This allows them to become more proficient in their growing knowledge of how to set up the investigation of manual-technical operations. They also record pre- and post-experimental data as they test the variable. After testing a gentle rain, they return to the baseline design of the model to test the effect of a heavy rain. This repetition results in the production of the journal entry seen in Figure 7.1 (p. 68), which shows the use of mathematical expressions*

*to label the stream table in precise detail, visual representations to show the way the model looked after each trial, and natural language to describe the results.*

## Rationale

In this chapter, we examine the role of **repeat** within the 5R Instructional Model. The critical aspect of this *R* is that it provides EMLs with repeated opportunities to engage with each mode of the hybrid language (natural language, mathematical expressions, visual representations, and manual-technical operations) to make meaning of concepts frequently encountered in the classroom. Students need to engage in scientific skills and processes and in the use of the hybrid language on multiple occasions to develop deeper conceptual understandings and language proficiency. Even though the idea of providing opportunities for repetition is simple and already used by many teachers, it is also complex and often not used to its full capacity. We speculate that this is the case because teachers are under great pressure to cover the curriculum, which allows few opportunities for repeated encounters with concepts and language.

Research in cognition provides evidence that task repetition is an important avenue for transferring new linguistic information to long-term memory, and it also supports the development of automaticity (Hattie and Yates 2014; Larsen-Freeman 2012). When readers are asked to perform the same reading task repeatedly, for example, they develop automaticity of lower-level tasks, such as sight word recognition and the ability to focus on language forms. Automaticity, in relation to repeated tasks, is also addressed in the literature addressing mathematics learning (National Council of Teachers of Mathematics 2014).

Lemke (1990) argues that students must have multiple opportunities to examine different ways to express the same thematic relationships. Repetition is not the parroting of identical expressions. Rather, teachers should use slightly different wordings to express the same relationships over the course of a lesson. Lemke refers to these repeated opportunities as "repetition with variation." He stresses that if students are to develop deeper understandings, they

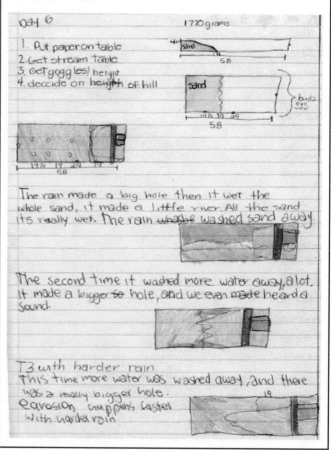

**Figure 7.1.** Student drawing of two views of the baseline followed by drawings of each trial

must not only listen to the teacher express these relationships but also engage in constructing their own meanings.

Returning to this chapter's scenario, we note that students were repeatedly asked to manipulate a variable as they engaged in three erosion experimental trials. The task slightly altered as students changed the manipulated variable each time. The investigation required students to set up the stream table several times, and with each trial, they repeated the manual-technical skill of assembling the baseline. Each repetition also required them to adapt their use of multimodal language to new situations.

We again take the opportunity to stress that the *R*s in this model do not occur in a specific order or in a consistent proportion. You have decisions to make during lesson planning (macro-scaffolding) and lesson execution (micro-scaffolding) about how and when to replace, reveal, repeat, and reposition multimodal language, concepts, and skills.

## Repeating the Hybrid Language of Science
### *Natural Language*

Larsen-Freeman (2012, 2014) introduces the notion of *iteration* as a way to further distinguish the traditional view of repetition (exact copying of language structures) from current views of repetition (providing learners with multiple opportunities to access concepts and language). She argues that the notion of iteration serves to recognize that although there is value in repeated activities, these activities must provide "learners with the opportunity to do something a little bit different each time they engage in a (repeated) particular activity" (2012, p. 204). In the opening scenario, we note how the repeated task allows students to extend their writing to include more complex syntactic elements as they compare observations. In the second and third trials, for example, they use expressions such as *more*, *bigger*, and *faster* to record more accurate observations and make comparisons across each repeated trial.

Similarly, Gibbons (2006, 2015) coined the term *message abundancy* to refer to the repeated use of patterns and routines that allows students to encounter an idea and notice how language can be used in different ways to express meaning—not just for the purpose of rehearsing language structures. In classrooms where students have access to message abundancy, "comprehension is also increased, because asking questions, exchanging information, and solving problems all provide a context where words are repeated, ideas are rephrased, problems are restated, and meanings are refined" (Gibbons 2015, p. 50).

As students engaged in their erosion investigations, Mrs. Malo repeatedly used the word *model* in a variety of ways:

> T: I want us to think about what we know about models. Earlier, we were talking about the 3-D models you made with clay. Let's look at the 3-D model that one group made. I can tell where you were on the campus from looking at this model. What is this a model of?
>
> S: The model is of the creek we visited.

T: You are correct. Scientists also make another kind of model. They make a model to see if they can answer their question. We made a model using the stream table to help answer our question about what effect the rain will have on the sandy hill.

Through this progression of teacher talk, the idea of a model as having more than one function was repeated. In the exchange, students revisit the notion of a three-dimensional model to think about how scientists use models to answer important questions. In addition, they had repeated manual-technical opportunities to create the model as they used the stream table to conduct their experiments (Figure 7.2). Mrs. Malo knew that a model was an enduring concept she would repeatedly use with other science topics throughout the year. Each opportunity to revisit a concept not only provides students with repeated hybrid language exposures within a particular context but also provides teachers with the opportunity to present the concept in a slightly more difficult context (Roessingh 2005).

Ultimately, students explored a variety of models (two-dimensional, three-dimensional, experimental, static, dynamic, and scaled). They even had fun contrasting their newly developed academic understandings of models with more familiar models they were likely to encounter in daily life (e.g., fashion model, role model). Multiple encounters with a concept throughout the academic year is a form of repetition often overlooked in science classrooms.

## Mathematical Expressions

In Chapter 6 (p. 57), we illustrated how Ms. Soto used an activity in which students measured their feet to reveal the ratio of foot length to height. Because scientists want to know a more accurate measurement, repeated measurements are needed. One way to get repeated measurements is for each student to conduct the investigation several times. In this situation, however,

**Figure 7.2.** Students measuring the height and length of the sandy hill to create an erosion model

the resulting data would be meaningless because students should get the same measurement with each trial. By contrast, repeated measurements that use class data (each student's result as a data point) provide information that gives a meaningful average. An average, as a mathematical expression, is important for meaning-making and can only be achieved with multiple iterations.

While Mrs. Malo, in the opening scenario, works with students to determine the effect of given variables on the yard, her colleague, Mrs. Malik in the mathematics department, is teaching scale factor. She does this by repeating activities requiring students to scale different objects mathematically. One way she provides repetition with variation is by having students read the trade book *Cut Down to Size at High Noon* (Sundby 2000). She uses this book because it introduces readers to a fictional town where two barbers use the mathematical concept of scaling to create haircuts. The author keeps the reader engaged while presenting accurate mathematics through natural language, mathematical expressions, and visual representations.

From this introduction to the concept of scaling, Mrs. Malik provides EMLs with repeated hands-on activities that require them to apply their growing knowledge and language of scaling through the use of centers. In center #1, students create scaled models of the Rocky Mountains; in center #2, they scale three-dimensional objects; in center #3, they scale recipes; and in center #4, they respond to the trade book. In each center, Mrs. Malik emphasizes mathematical expressions and requires students to use all modes of the hybrid language to demonstrate their understanding of scale factor. Ultimately, students will develop a two-dimensional scaled model using the stream tables from Mrs. Malo's science class. This collaboration between disciplines provides the additional value of helping science students construct explanations of a natural phenomenon by using mathematical and computational thinking based on a physical model (as stressed in the *Next Generation Science Standards* [NGSS Lead States 2013]).

## Visual Representations

In the science classroom, there are many opportunities for students to engage with visual representations. Mrs. Malo takes the time to show students how to read different visual representations and explains different ways to represent ideas visually. She repeatedly uses the science textbook to help students recognize that visuals are oriented from different perspectives (e.g., front, side, lateral, bird's-eye, worm's-eye) and sizes (e.g., microscopic, actual, macroscopic).

Throughout Mrs. Malo's erosion lesson, students have repeated opportunities to create visual representations of the stream table. Take another look at Figure 7.1 (p. 68). The first two drawings on the right show the baseline with measurements for the slope (hill). Each representation varies in terms of the information it communicates. The worm's-eye view (side view) highlights the height of the hill on the model. The bird's-eye view (top view) shows how far along the stream table to position the sand. The next drawing shows the setup just prior to pouring water into the rain cloud.

Students repeatedly create a new representation after each of the three trials to denote how the sand moves across the model after it rains. For example, students repeat a drawing from their journals (Figure 7.3, p. 72), and this repetition provides the opportunity not only to discuss the drawing but also to edit and enhance it.

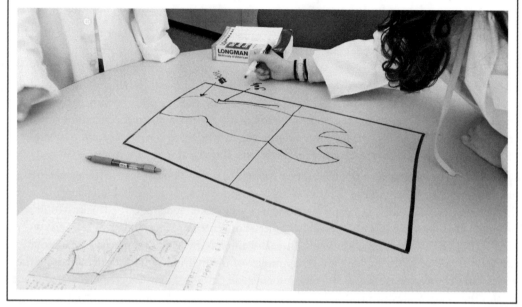

**Figure 7.3.** Students repeating the drawing with the addition of new information

Students also repeatedly prepared data tables in which they recorded the manipulated variable. The first data table (wind) was developed with extensive scaffolding from Mrs. Malo with explanations about what the rows and columns meant. As students tested another variable (gentle rain), they constructed a similar data table with less help from their teacher. By the third variable tested (hard rain), students were able to construct the data table with very little help.

## Manual-Technical Operations

Science is a social activity and as such involves movement at various levels. Classrooms where EMLs are engaged in inquiry-based instruction establish an environment in which there is codependent emergence of language, material activities, and equipment manipulation. Therefore, repeating ideas and skills in science is necessary if deep understanding is to occur. Students in Mrs. Malo's class repeatedly used the model of a stream table while they studied erosion. They repeated the experience of setting up the original design to test new variables. Each time the stream table model was assembled, students were reminded to repeat the established baseline. As expected, students became very good at setting up the model and testing another variable. Through the repeated manipulation of objects, they communicated about erosion even as they developed the natural language expected for class participation.

In Mr. Jimenez's class (Chapter 5, p. 45), EMLs used a model of a wind turbine to test variables (e.g., type of blade, number of blades, degree of tilt, placement of blades). Each time they assembled the basic wind turbine and connected the voltmeter, they became more literate in the manual-technical operation. Not only were students repeating the manual-technical skill of preparing the wind turbine, they were also repeating the conceptual understandings of a dynamic model, fair test, and variable manipulation.

Another example of a teacher using repeat can be seen in Ms. Soto's CSI unit (Chapter 2, p. 9) as EMLs studied DNA and genes. Students were given one known and four unknown samples of DNA, and they used gel electrophoresis to separate the fragments as they learned about the unique qualities of DNA. On their first attempt, students had difficulty using a pipette (Figure 7.4) to put a DNA sample into the well of an electrophoresis chamber. With repeated attempts, though, they became quite good at moving the sample from the vial to the pipette to the well.

The manipulation of scientific equipment is important in meaning-making, and each repetition helps concretize the experience. Setting up equipment multiple times allows EMLs to focus on a concept, and the repeated nature of this experience allows students to use manual-technical operations, natural language, and mathematical expressions. When students return to their seats to record their data, they add another mode—visual representations. This helps them develop and express a more complex conceptual understanding of the science concepts. As conceptual understanding increases, the use of multimodal language also increases.

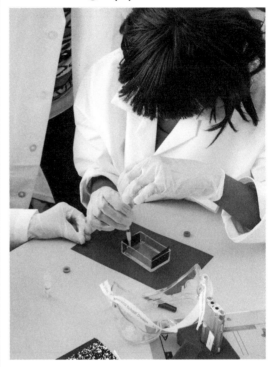

**Figure 7.4.** Student moving sample to well using a pipette

As seen in the examples above, science classrooms can be a place where students learn their "way around the laboratory [to become] familiar enough with equipment to 'figure it out'" (Hunter, Laursen, and Seymour 2007, p. 65). Labs should provide EMLs with access to many diverse tools (natural language and equipment) and with multiple opportunities for feedback on their appropriate use (Richmond and Kurth 1999).

## Concluding Remarks

Repetition is used to provide EMLs with multiple opportunities to encounter and express scientific meaning using all of the modes of the hybrid language. Rather than the exact duplication of a specific way of expressing meaning, this *R* provides EMLs with slight variations in ways to examine and express similar meanings within the inquiry-based classroom.

# Chapter 8

# Reposition

D r. Enuge wants to stress the crosscutting concept of energy, the physical science content of energy, and the practices of science. To develop the discourse of science, Dr. Enuge provides emergent multilingual learners (EMLs) with opportunities to refine their ability to construct and refine meaning within one mode and across all modes of the hybrid language.

*EMLs are investigating the best design for wind turbines. They know the target design produces a large output of energy (electricity) over a sustained period. The science lesson has been going on for some time when Dr. Enuge asks the class to discuss what they do each time they conduct a trial. Miguel says they change some things about the wind turbine model. With more prompting, he explains that something is changed each day and they use a chart to record the results.*

*Dr. Enuge asks students to examine their journals and tell her more about the changes they observed. After looking at a chart in her journal, Erika says that on the first day, they changed the number of blades, and on a different day, they changed the tilt of the blades. Erika also reminds the class that they completed more than one trial, calculated the average, and recorded information in a data table.*

*Dr. Enuge says, "When we talked about the investigations, we said that we change the variable. I want us to use a more academic word instead of change. The new word is* manipulate, *so we call this variable the manipulated variable. A manipulated variable is the variable we select to change." She purposefully replaces a common word used by students with an academic word and provides support for more formal language that resembles written text. She also writes the following sentence stem on the board: "Our manipulated variable is _____."*

Dr. Enuge asks students to return to the chart and complete this sentence with the variables they had changed. Carlos responds by saying, "Our manipulated variable is the number of blades."

Dr. Enuge points to the word manipulated on the board and asks students if they have seen this or a similar word before. One student shares how her mother uses a similar word in Spanish—manipular—when talking about one of the bossy girls on her soccer team. This comment prompts Maria to comment how manipulate reminds her of the word mano (hand) in Spanish. The class then discusses how this notion relates to their use of manipulated variables (i.e., they changed one variable with their hands).

Dr. Enuge has students focus on the effect of the manipulated variable on the wind turbine model (Figure 8.1) and says, "There is another name for the variable where change happens as a result of our manipulation of the investigation. Each variable is tested to see if the voltage will change. Scientists would say the variable that results is called the responding variable. Let's put that sentence on the board." She then writes a new sentence frame on the board: "Our responding variable is _____."

The class again engages in language talk and discusses why the word responding serves to describe this type of variable. Dr. Enuge asks students to identify a manipulated variable. They respond by telling her they are testing the position of the blade on the wind turbine. Students reposition their language from "We change the position and got some energy" to "Our manipulated variable is the position of the blades, and our responding variable is the amount of energy produced." As the lesson progresses and more tests are conducted, Dr. Enuge continues to prompt students to use the new sentence stems to explain their investigations.

**Figure 8.1.** Whiteboard showing sentence stems for types of variables

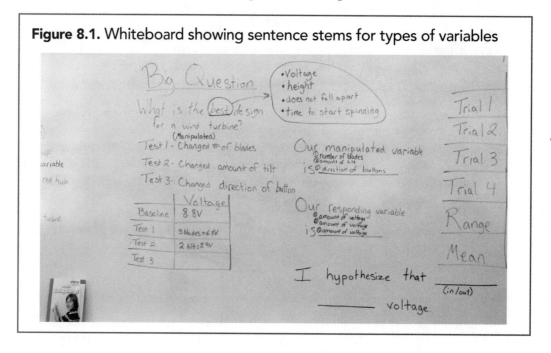

## Rationale

In this chapter, we introduce **reposition** as a component of the five *R*s. The critical aspect of repositioning is to provide EMLs with extended opportunities to assume the complexities of the hybrid language. As you create opportunities to reposition, EMLs develop increasing sophistication in their ability to create meaning through and across each of the hybrid language modes (natural language, mathematical expressions, visual representations, manual-technical operations). This in turn helps them better understand the science concepts developed in your classroom.

Repositioning occurs when students have multiple experiences with the science concepts in a lesson. Prior to the discussion highlighted in the opening scenario, EMLs engaged in manual-technical operations as they assembled repeated trials in which they manipulated a variable on their wind turbine model (e.g., they changed the number or tilt of the blades). After observing and measuring the results of each change, they recorded their observations on a class-generated chart.

To be recognized as members of the discourse community of Dr. Enuge's science classroom, EMLs learned to coordinate their communication across the modes of the hybrid language. In the context of these interactions, Erika used a data table created by the class for each trial. To report her findings, she focused on three modes of the hybrid language—visual representations, mathematical expressions, and natural language. Though the modes each represent different ways of expressing meaning, they do not communicate the whole event in isolation. Erika had to coordinate across modes to process and express her understanding of the investigation.

From the discussion, Dr. Enuge had evidence that students understood the concept. She was also aware that EMLs needed additional support to communicate their understanding of the relationships between and among the variables they were observing. She scaffolded this more complex language to help students communicate the relationships across variables by providing them with two sentence stems:

Our manipulated variable is _____.

Our responding variable is _____.

In repositioning the language to express this relationship, Dr. Enuge demonstrates an awareness that this type of reasoning is constructed with language and is essential to doing science. By drawing students' attention to ways of communicating the observed scientific relationships, Dr. Enuge makes language visible to the EMLs in her classroom. Using language to talk about language (i.e., metalanguage) in the science classroom provides EMLs with a tool to reflect on their own use of the hybrid language.

Ultimately, repositioning supports EMLs as they develop the ability to coordinate complex meaning across the hybrid language. Rather than learning each mode as a discrete skill, EMLs benefit from extensive opportunities to engage in meaning-making across all modes. They also benefit from opportunities to understand how each mode serves a different purpose yet comes together when communicating meaning. As with other *R*s, repositioning may be planned (macro-scaffolding), or it may occur during instruction (micro-scaffolding).

# Repositioning the Hybrid Language of Science
## Natural Language

To develop the language of the science classroom, teachers must support students as they move from informal (spoken-like) to formal (written-like) language. We highlight how teachers support EMLs repositioning language to emphasize text genres and linguistic features typical of the written forms of science. In this section, we feature two examples of repositioning as a student appropriates text genres and grammatical structures within an erosion investigation.

Bernardo has been in the United States for less than a year. In a journal entry (Figure 8.2 and Table 8.1), he responds to a "What I Did" / "What I Learned" reflection in Dr. Molly's class. The "What I Did" component of the journal entry is an example of a personal recount where students draw on their own experiences to retell the events that occurred in class on the previous day. Bernardo organizes the series of events using words that signal a sequence: *first, second, next,* and *after.* He also uses present tense to express what happened and includes some specific science words related to the unit.

In the "What I Learned" section of the journal entry, Bernardo engages in an explanation of the events he previously outlined. In this entry, he uses signal words that indicate logical relation-

**Figure 8.2.** Repositioning language journal entry: "What I Did" / "What I Learned"

| What I did | What I learned |
|---|---|
| First we chec and put the new words in the Academic or informal Second we write about the book named Erosion, we write about the types of erosion an ~~the~~ the cause, next we aunser the question of the what will happen to Dr. Molly's yard if we get a gentle rain, next we ~~make one mode~~ make one mode of the rain, after we write about the ~~chang~~ change | I leanned mor words one example is: Erosion, Decrease, increase and proportion after we learned mor about the erosia how many types of erosion have, what is the cause, later we learned more about the Erosion by water in the model of Dr. Molly's yard after write about the cause of Erosion by water, an learned this Erosion. 1. water, ice and wind 2 change the earth surface *Can you tell me what happens* |

ships typical of an explanation. For example, he establishes a taxonomy that signals classification (e.g., *one example is, many types*). He also uses words that signal a cause-and-effect relationship (e.g., *what is the cause, the cause of*) in his attempt to explain his current understanding of erosion.

**Table 8.1.** Transcription of Bernardo's Journal Entry: Personal Recount and Explanation

| Step | What I Did [Personal Recount] | What I Learned [Explanation] |
|---|---|---|
| 1 | **First** we chec [check] and put the new words in the ~~formal~~ Academic or informal. | I learned **mor** [more] words **one example is:** Erosion, Decrease, increase and proportion. |
| 2 | **Second** we write about the book named Erosion. We write about the types of erosion an the cause. | **After** we learned mor about the eresio [erosion] **how many types** of erosion have. **What is the cause.** |
| 3 | **Next** we aunser [answer] the question of the what will happen to Dr. Molly's yard if we get a gentle rain, **next** we ~~produce the~~ make one modle [model] of the rain. | **Later** we learned more about the Erosion by water in the model of Dr. Molly's yard. |
| 4 | **After** we write about the change. | **After** write about the **cause of Erosion** by water. An learned this Erosion: 1. water, ice and wind 2. change the earth surface. |

Repositioning occurs as EMLs have experiences with science content and related language. To contextualize the previous personal recount and explanation, Bernardo participated in the following language experiences:

1. **Word Wall:** Bernardo's teacher used a word wall to reload important words and concepts from previous lessons. In his journal, Bernardo alludes to a reloading activity in which the teacher asked students to consider whether the words from previous lessons were used in informal or academic settings. In his explanation, Bernardo highlights that he learned new words and identifies four of those words: *erosion*, *decrease*, *increase*, and *proportion*. We will further discuss reloading in Chapter 9 (p. 89).

2. **Book Clubs:** On the previous day, Bernardo participated in a book club and wrote a response about the book he read. His teacher uses book clubs to engage EMLs in readings that extend their conceptual and linguistic understandings of erosion beyond inquiry-based lessons. Bernardo had self-selected a book from an erosion text set. The book clubs met in small groups to read and discuss selected chapters of the book. Following the reading, Bernardo responded to the book he had read. In the "What I Learned" section of his journal entry, Bernardo demonstrates his emerging understanding of the causal relationship between natural forces and erosion.

The teacher further supported Bernardo's linguistic development by helping him understand how written language is typically organized in the science classroom. In the following journal entry, which he wrote three days after the entry previously discussed, Bernardo repositions his earlier understanding of erosion (Figure 8.3 and Table 8.2). In this repositioned definition of erosion, Bernardo uses connectors (*cause, after, and, so*) to signal the relationship between natural forces (water, wind, and ice) and the movement of earth materials signaling erosion.

Not evident in Bernado's journal entries is the role the teacher's scaffolding played in his linguistic development. His second journal entry was written after the teacher engaged the class in an analysis of how authors craft formal definitions. Knowing students were comfortable with the concept of erosion, the teacher wanted them to communicate

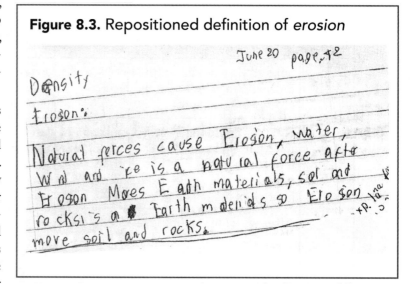

**Figure 8.3.** Repositioned definition of *erosion*

their understandings using more formal, written-like language. EMLs orally reviewed their own understanding of erosion and examined a formal definition (which led to the development of Bernardo's definition): "Erosion is the action of natural forces, such as water, wind, or ice, that remove soil from one location to another on Earth's crust."

**Table 8.2.** Transcription of Bernardo's Journal Entry: Repositioned Definition of *Erosion*

Density
Erosion:
Natural forces **cause** Erosion, water, wind and ice is a natural force **after** Erosion moves Earth materials, soil **and** rock is a Earth materials **so** Erosion move soil and rocks.

As part of this discussion, the teacher leveraged a language moment (Heritage et al. 2015) and called students' attention to the density of the definition by discussing the amount of information packed into one sentence. This type of information packing is formally known as *nominalization* (Fang 2006; Fang et al. 2010; Gibbons 2009; Schleppegrell 2004), which refers to the process of changing verbs, adjectives, or clauses into nouns (e.g., "Erosion is the action of natural forces such as water, wind, and ice that move soil from one location to another on the Earth's crust" can be nominalized into the word *erosion*). Nominalizations can challenge EMLs because, when unpacking a great deal of information, they run the risk of losing the text's meaning.

To scaffold the linguistic concept of density, the class first examined the use of population density, which was a more familiar concept for students. Then they examined how the concept of density is communicated through other forms of the hybrid language (visual representations and mathematical expressions) (Figure 8.4).

Using these notions of density, the class examined a definition of *erosion* and discussed how its author had compressed information into one sentence. They circled the multiple ideas found in the sentence (Figure 8.5).

During the discussion, students also identified how the author used connectors (e.g., *such as, or, that*) to link ideas. At that point in the lesson, the teacher handed out paper strips with the following sentences and asked students to work in pairs to manipulate the strips and come up with their own definition of erosion.

1. Erosion is an action of natural forces.

2. Soil is an Earth material.

3. Erosion moves soil.

4. Natural forces cause erosion.

5. Water is a natural force.

6. Wind is a natural force.

7. Ice is a natural force.

Students shared several ways they had linked the sentences to produce various definitions of *erosion*. The teacher discussed how packing information into dense sentences helps authors communicate scientific information concisely. The activity culminated as students wrote their own definitions, and Bernardo's journal entry represents his effort to refine written language to come up with a definition.

It's important to note here that even though the language of the science classroom is characterized by the use of particular grammar structures that often make it denser, the teacher did not introduce nominalization as an isolated linguistic feature. Instead, the class investigated ways grammatical functions support meaning-making.

**Figure 8.4.** Density: Visual and mathematical expression

Density in Science

Density in Mathematics

Mathematical Sentence

$(3+3+3)+(4+4+4+4)+2=$

$(3\times3)+(4\times4)+2=$

$(3)^2+(4)^2+2=$

*Note:* Adapted from Kucer and Silva 2013.

**Figure 8.5.** Density: Natural language

Erosion is the (action) of (natural forces) such as (water,) (wind,) or (ice) that (move) (soil) from one (location) (to another) on the (Earth's crust.)

To support students like Bernardo in refining their use of written text genres, teachers are well aware that growth is gradual and comes as a result of repeated opportunities to engage in making sense of natural phenomena. In this sense, we concur with Heritage and colleagues (2015) who remind us that "language use spirals in sophistication, depth, and eventually correctness, based on the opportunities students have to use it to express important ideas, and it always develops simultaneously with conceptual understanding and analytical disciplinary practices" (p. 38). As Bernardo continued to engage in meaning-making within the classroom, the teacher continued to scaffold the development of language he needed to fully participate in the discourse of the community.

Fortunately, the publication of the *National Science Education Standards* (National Research Council 1996) created an interest in helping *all* learners engage in meaning-making practices to make sense of scientific phenomena. This interest resulted in several publications we view as critical in helping science teachers reposition language. Hand and his colleagues (Hand and Keys 1999; Keys et al. 1999) developed the Science Writing Heuristic, which has been instrumental for scaffolding students' science knowledge construction and communication. Later, the *Next Generation Science Standards* (NGSS Lead States 2013) provided further guidance on science instruction for diverse learners. Teachers have also benefited from National Science Teachers Association and National Council of Teachers of Mathematics publications and professional development to support the development of student-centered classroom discourse (e.g., Cartier et al. 2013).

## Mathematical Expressions

When we reposition mathematical expressions, we help students develop the language necessary to communicate mathematical understandings. This includes the use of precise nomenclature of numbers (values), formulas, and related mathematical concepts. An example occurred in Dr. Enuge's class when her students were discussing the voltage readings from their wind turbine investigation (Figure 8.1, p. 76). One student reported that his voltage reading was "seven point one." Another student reminded him that the word *point* does not indicate the place value, and it should be read "seven and one-tenth":

S1: Our blades turned fast.

S2: Yes, we got seven point one.

S1: What?

S3: No, not seven point one; you should say seven and one-tenth.

S2: Oh, that's right. Seven and one-tenth. We got seven and one-tenth.

S4: What about seven and one-tenth?

S1: Volts! Seven and one-tenth volts!

S2: Yes, we got seven and one-tenth volts.

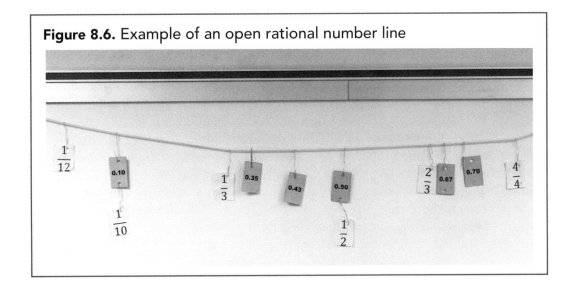

**Figure 8.6.** Example of an open rational number line

These same students had studied place value and decimals during their mathematics class to better understand the reading on the voltmeter. In other words, they were developing the concept of place value as it pertained to digits found to the right of the decimal point.

As a teacher, you should note that saying "My reading is seven point one" may indicate a lack of understanding of the value. Students should be reminded that language carries conceptual meaning. A repositioning of this statement to "My reading is seven and one-tenth" reflects the conceptual understanding of place value needed to support the discourse within the science classroom.

During mathematics instruction, Dr. Enuge had students reposition mathematical expressions with an open rational number line (Pearn 2007). This type of number line includes small movable cards that depict mathematical expressions as fractions, decimals, and percentages between 0 and 1 (Figure 8.6). As students place more cards on the number line, numbers are repositioned to show a more precise representation of their value. At the same time, students can see the equivalency of numbers. For example, the value of 1 can be written as 4/4, 17/17, or any fraction where the numerator and denominator are equal. Although this may seem trivial for the EMLs in Dr. Enuge's class, a mathematical understanding of repositioning the value of 1, which can be written an infinite number of ways, is essential to the Identity Property of Multiplication.

Another way to use equivalency on the number line is by focusing on the relationship between fractions, decimals, and percentages. For example, students should recognize that ½ is equivalent to both 0.5 and 50%. This mathematical repositioning is used in science class and in students' everyday lives. In the science classroom, the importance of learning this equivalency is evident when expressing measurements. For example, scientists say "The length of the object is a half-meter" but not "The length of the object is 50 percent of a meter." Likewise, we would say "About 50 percent of the forest is depleted" but not "About 0.5 of the forest is depleted."

A final example of repositioning occurs when helping high school students use denser mathematical expressions to express mathematical meaning more efficiently. While studying mitosis, students learn that the process adds cells to a multicellular organism. Each cell division doubles the number of cells. To calculate the total number of cells, students begin with addition. This works up to the point where the string of addition is too tedious to express in oral or written form (column 2 of Table 8.3). Students need to reposition from addition to a formula for calculating the number of cells in an organism given the number of times that cell division has occurred. The mathematics involved in determining the number of cells created from one cell after *n* splits can be solved by building a table and looking for patterns. Students must look for patterns and reposition the addition into a multiplication problem (column 3 of Table 8.3). If a large random number is chosen, calculating the number of new cells using addition or multiplication would be time-consuming and inefficient. Therefore, as students examine the pattern that occurs in multiplication, the teacher can help them reposition the multiplication expression into a general algebraic expression of exponential growth (column 4 of Table 8.3).

**Table 8.3.** Example of Repositioning From Addition to Multiplication to Exponential

| Number of Cell Divisions | Addition | Multiplication | Exponential | Number of Cells |
|---|---|---|---|---|
| 0 | | | $2^0$ | 1 |
| 1 | $1 + 1 =$ | $1 \times 2$ or $2 \times 1$ | $2^1$ | 2 |
| 2 | $2 + 2 =$ | $2 \times 2$ | $2^2$ | 4 |
| 3 | $4 + 4 = (2 + 2) + (2 + 2)$ | $2 \times 4 = 2 \times 2 \times 2$ | $2^3$ | 8 |
| 4 | $8 + 8 = (2 + 2 + 2 + 2) + (2 + 2 + 2 + 2)$ | $2 \times 8 = 2 \times 2 \times 2 \times 2$ | $2^4$ | 16 |
| 5 | $16 + 16 = (2 + 2 + 2 + 2 + 2 + 2 + 2 + 2) + (2 + 2 + 2 + 2 + 2 + 2 + 2 + 2)$ | $2 \times 16 = 2 \times 2 \times 2 \times 2 \times 2$ | $2^5$ | 32 |
| 10 | $512 + 512 =$ | $2 \times 512 =$ | $2^{10}$ | 1,024 |
| 25 | $16,777,216 + 16,777,216 =$ | $2 \times 16,777,216 =$ | $2^{25}$ | 33,554,432 |
| 47 | ???? | ???? | ???? | ???? |

When repositioning language, teachers provide EMLs with opportunities for metalinguistic reflection. The following journal entry is an example of a student's reflection following a language moment in which the class had an explicit discussion on the ways mathematics supports communication within the science classroom (Figure 8.7). In the entry, the student responds to this prompt: "As a scientist, how do I communicate mathematically?" Here, the student references the measurements the class took to set up a model they used for an erosion investigation (Table 8.4).

**Table 8.4.** Transcription of Journal Entry

As a scientist I communicate by using math tools. For example, in Mrs. _____ class, my table and I were measuring the height of the sand and how deep, model.

In addition to using natural language, this student integrated visual representations to demonstrate how she used mathematics to measure the height and depth of the stream table and the weight of the sand she would use as a baseline for the investigation.

Let us again emphasize that to support EMLs as they develop the complexities of the hybrid language, they need to have extensive opportunities not only to engage with the various modes of the language but also to examine how these modes interact in meaning-making.

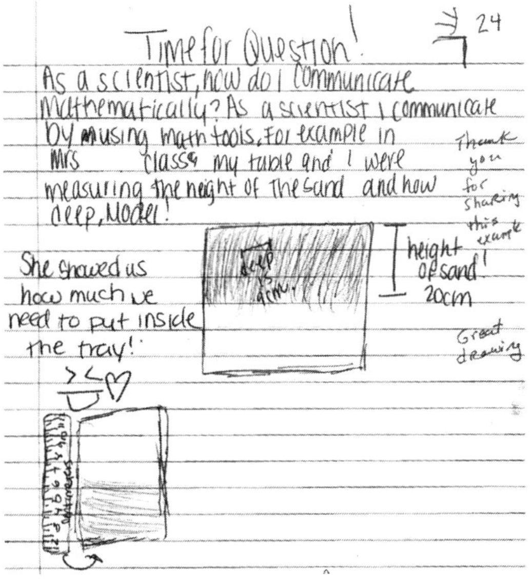

### Visual Representations

When working with EMLs, we seek opportunities for students to interpret and create science data through the use of multiple visual representations. As you may recall from Chapter 2 (p. 9), visual representations can vary in terms of complexity to serve different functions. For example, students often find creating a T-chart to be a simplistic but highly effective recording device for data collection (Table 8.5). This visual only allows one interpretation of the data; however, students are often asked to communicate the data as a graph.

Going from a chart to a graph requires the repositioning of data to a visual representation and necessitates interpretation. A graph requires students to collapse the four trials into one visual (Figure 8.8, p. 86). When looking at a graph, students must know what information goes on both the *X*-axis and the *Y*-axis. The manipulated variable (height, in this case) is placed on the *X*-axis, and the responding variable (rebound, in this case) is placed on the *Y*-axis.

**Figure 8.7.** Journal reflection: As a scientist, how do I communicate mathematically?

**Table 8.5.** T-Chart to Collect Data From Four Trials

| Trial | Height of Drop (MV) | Length of Rebound (RV) |
|---|---|---|
| 1 | 1 meter | 9 centimeters |
| 2 | 2 meters | 14 centimeters |
| 3 | 3 meters | 21 centimeters |
| 4 | 5 meters | 23 centimeters |

In addition, students must know that the distance may appear to change along the *X*- and *Y*-axes, depending on how the information is positioned. For example, students could use a scale of 1 (e.g., 1, 2, 3, 4 . . . 23) or 2 (e.g., 2, 4, 6, 8 . . . 24) or any other fixed proportional spacing, which would change the slope of the line. When you reposition from one visual to another, you facilitate student movement into the discourse of science.

As with the previous example, Figures 8.9 and 8.10 are examples of EMLs' journal pages showing visual representations of data. To reinforce the process of investigation, students were asked to investigate the languages they used at home. They identified the questions, designed the data collection, and prepared methods of data recording. Figure 8.9, a bar graph, represents data about the various languages spoken by class members. Figure 8.10, a Venn diagram, displays the linguistic relationships among the data set. In this case, the second representation is more complex and goes beyond the typical Venn diagram used in most classrooms to examine relationships. The diagram allows students not only to determine how many family members speak each language but also to examine the intersection among the various languages.

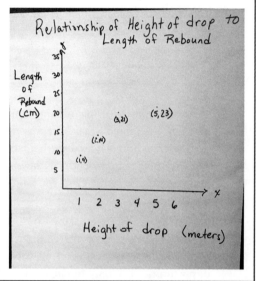

**Figure 8.8.** Repositioning a data table to a visual representation

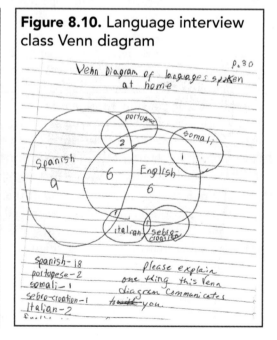

**Figure 8.9.** Language interview class bar graph

**Figure 8.10.** Language interview class Venn diagram

## Manual-Technical Operations

Using tools and body movements to make meaning of natural phenomena begins with simple tools and movements. Part of repositioning happens naturally as students progress from elementary to high school. However, the teacher must facilitate repositioning by making tools available to students. One example of tool repositioning familiar to many science teachers is when students move from using a hand lens to a microscope.

Another example of repositioning occurred in Mrs. Malo's class (Chapter 7, p. 67) during the erosion unit. Students began a lesson by weighing sand to establish a baseline model for further variable manipulation. Students already understood the concept of weighing from both personal and informal use of bath scales at home and two-pan balances found in elementary classrooms. The two-pan balance provided them with a context-rich tool that explicitly showed both sides must equal or balance. As manual dexterity increased and more precise measurements were needed, student engagement with tools was repositioned to the triple-beam balance, which is a more sophisticated weighing tool. This tool (which was less familiar to students and had fewer contextual clues) uses counterweights along three arms to measure the weight of an item. Further repositioning was seen as students moved to digital scales—where the physical process of balancing is lost but the information is highly accurate (Figure 8.11).

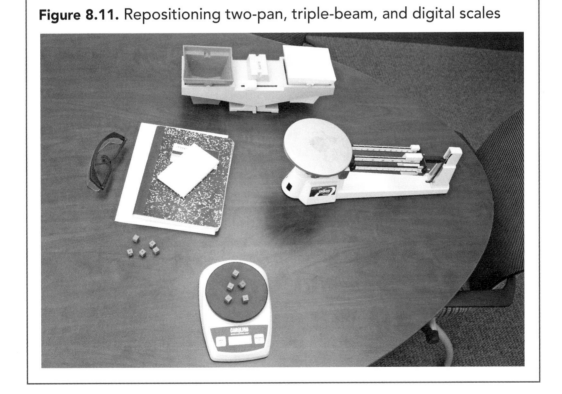

**Figure 8.11.** Repositioning two-pan, triple-beam, and digital scales

## Concluding Remarks

Repositioning moves EMLs into the discourse of the scientific community. To talk like, act like, and look like scientists, students must be engaged in fully using the meaning-making affordance of the hybrid language. Reposition lessons not only provide students with opportunities to assume the complexities of scientific language but also make it visible by engaging students in talk about hybrid language.

# Chapter 9

# Reload

M s. Blair is aware of the need to provide emergent multilingual learners (EMLs) with multiple opportunities to focus on word meaning and usage. In her classroom, she stresses mathematics, computational thinking, and communicating information. We enter her classroom as students review and make connections across words previously encountered in a graphing lesson.

Students are working at tables in groups of four. At the beginning of the lesson, they are asked to go to the word wall and select one of the words. They are to identify and discuss connections among the four words. As students discuss the meanings of their words, one states, "I have *horizontal*, and I think it should go with *vertical*." Another responds, "No! It goes with x-*axis*." "But what about y-*axis*?" asks a third student. The fourth student pipes up, "I think they all go together."

One student says the word *horizon* and is corrected by a tablemate who thinks that if you say it that way, it means something else. They call Ms. Blair over to clarify the differences between *horizon* and *horizontal*. She helps students consider how both words have a common root—*horizon*. Together, they examine how the suffix *-al* changes the meaning. The students then decide they want to add *-al* to the ongoing chart the class uses to identify common prefixes and suffixes they encounter in science. Maria, one of the Spanish speakers in the group, recognizes *horizontal* as a Spanish cognate. As is common practice in the class, Maria asks Ms. Blair if she can add a green sticker to the word *horizontal* to signal that it is a cognate before placing it back on the word wall.

By the end of the discussion, students in one group begin to use a classmate's arms as a place to display the words (Figure 9.1, p. 90). When they are ready to present their findings, students in this group stand between and among the words *horizontal*, *vertical*, x-*axis*, and y-*axis* by positioning

the student's arms to represent the perpendicular lines of a coordinate plane. Ms. Blair and the students then revisit the words on the word wall, linking the discussion to concepts that were introduced. Having overheard her group's discussion, Ms. Blair calls on Maria to add more information. Maria informs the class that *horizontal* is a cognate and explains her thinking. Ms. Blair concludes this lesson component by directing students to examine their journals for evidence of use of these words. She then supports the class in coming up with definitions based on their usage.

## Rationale

In this chapter, we discuss the *R* of **reload**. The critical aspect of reloading is that it provides EMLs with opportunities to revisit words they have previously encountered within a lesson. Reload differs from the *R*s introduced in the preceding four chapters (replace, reveal, repeat, and reposition) in that it is context-reflective and focuses on one component of natural language: vocabulary. Though we have stressed the need to anchor language learning within the context of the lesson, when reloading, we provide EMLs with opportunities to reflect on vocabulary that first emerges within science lessons. As students become more comfortable with concepts and words, activities can offer less contextual support. Reloading vocabulary after the lesson (e.g., at the beginning of the next lesson or as part of interdisciplinary work during mathematics or language arts) is essential for building academic success in science classrooms.

**Figure 9.1.** Enactment of the word relationship among x-*axis*, y-*axis*, *vertical*, and *horizontal*

In the opening scenario, students were engrossed in a discussion about the meaning of *horizontal*, *vertical*, x-*axis*, and y-*axis*. They examined the relationships among these words, which were first introduced the previous day within a context anchored lesson on graphing. They were then reloaded by Ms. Blair as part of a context-reflective activity. Ms. Blair captured the words on the word wall so students could return to them later. This reloading lesson provided EMLs with the opportunity to further manipulate word meaning and usage.

When reloading words, it's important to use the many strategies practiced in the classroom to support vocabulary development. Rather than front-loading vocabulary by preteaching word meanings, highlight vocabulary only after students first encounter it within the context of a science lesson. Because we believe that inquiry-based science should begin with an examination of natural phenomena (as discussed in Chapter 6, p. 57), front-loading vocabulary prior to the science lesson becomes counterproductive. Within the 5R Instructional Model, reloading lessons foster word consciousness and support word learning strategies (Graves et al. 2013; Stahl and Nagy 2006). Reloading also develops a general interest in and an awareness of words—as well as the metalinguistic knowledge needed to know how these words are used.

## Reloading in the Science Classroom

The importance of vocabulary knowledge—particularly in middle and secondary schools—is widely documented in the research literature (e.g., Carlo et al. 2004; Graves et al. 2013). Ms. Blair is aware of the large number of new words EMLs encounter in her classrooms. In fact, the number of new science words students encounter in middle and secondary school textbooks is greater than the new vocabulary expected in a foreign language course (Groves 1995; Yager et al. 2009). Science is often the context in which many new words are introduced, and students' achievement—as well as your own success as a science teacher—is assessed based on their competence with these new words (Yager et al. 2009).

### *Acquiring Different Types of Words*

In science classrooms, EMLs encounter technical and abstract words and concepts that differ from those they use in everyday interactions (Fang and Schleppegrell 2008). Students might talk about *rain* at home, but in the classroom, they encounter *precipitation*—a term that implies a deeper understanding (i.e., that as the atmosphere becomes saturated with water vapor, liquid water is released from clouds). In the science classroom, common English words can also have specialized meanings (e.g., *work, force*). However, not every word that students encounter is content related. When EMLs develop vocabulary, they learn different categories of words. Beck and McKeown (1985) conceptualize vocabulary learning in terms of a three-tier instructional framework.

Tier 1 words include the most basic words (e.g., *animal, house, eat, run*) needed for communication in everyday contexts. Like native English speakers, EMLs learn these words in their first language. Unlike native English speakers, however, they might not be familiar with the wide range of English words that are categorized within Tier 1. Because these words are so familiar to you, their meaning may seem self-evident, and a short explanation may be necessary.

Tier 2 words frequently appear in written texts and tend to be common in adults' vocabulary yet may be unfamiliar to EMLs. These include words found in more than one discipline. For example, the word *predict*, though used in both English language arts and science classrooms, carries slightly different meanings and usage. In English class, students are encouraged to use context and background knowledge to make reading predictions about what will happen. In science, by contrast, students are expected to make predictions and develop an outcome statement based on data. Tier 2 vocabulary can also include phrase clusters (e.g., *make a claim*) and signal words (e.g., *because, since, if . . . then*), which serve to express relationships and are closely associated with the factual and analytical genres highlighted in the *Next Generation Science Standards*.

Tier 3 words convey very specific, often technical meanings. They have low-frequency use and occur most commonly in textbooks and other forms of formal language. They are critical to scientific language because they allow scientists to establish classes, categories, and taxonomic relationships that cannot be replaced by other, more common words (Fang 2006). Neither native English speakers nor EMLs are likely to know Tier 3 words when they first encounter them. These words are frequently based on Latin and Greek roots and contain more than one morpheme, which can change meaning—as we saw in the scenario when Ms. Blair and her students examined the words *horizon* and *horizontal*.

In the following sections, we highlight different ways in which students interact with words during reloading lessons. You will likely recognize that the strategies we use are commonly used in many classrooms to support vocabulary development. Our aim is to highlight methods to help EMLs focus on word relationships, meanings, and parts. We do not seek to provide a comprehensive description of vocabulary strategies.

## Word Walls

Ms. Blair consistently relies on a word wall to display words that she and her students consider important. She uses the word wall to create awareness of words and the metalinguistic knowledge needed to know how words are used. Her students often express that the reason they find science words difficult is because they are learning English. Through reloading lessons, Ms. Blair helps EMLs understand that when talking science, they are engaging in a specialized register that is no more difficult than any other (Lemke 1990).

As vocabulary emerges within the context of science lessons, Ms. Blair and her students work together to decide which words need to be displayed on the word wall. This decision may be based on (1) predetermined (macro-scaffolding) Tier 2 and Tier 3 words she views as essential to the unit of study, (2) formative assessment (micro-scaffolding) of Tier 1 and Tier 2 words during instruction, or (3) students' spontaneous (micro-scaffolding) identification of interesting words they would like to add to the word wall. Identified words are written on strips of paper and displayed on the word wall (Figure 9.2).

## Word Relationships

One way to reload vocabulary is through lessons that provide students with opportunities to understand how science words relate to one another. Scientific meanings, as we discussed in Chapter 6, are best expressed in terms of thematic patterns. To communicate in science class,

**Figure 9.2.** Word wall in a science classroom

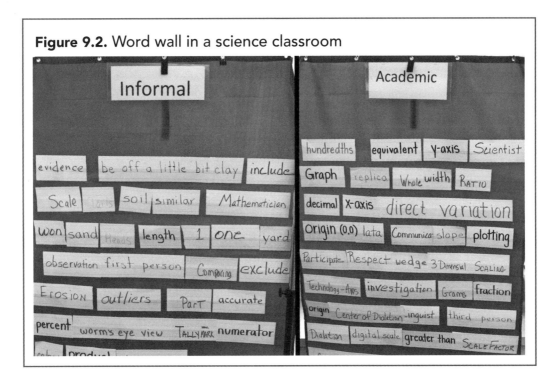

students must go beyond learning basic word definitions. They need to learn how to use terms in relation to one another. In the opening scenario, Ms. Blair intentionally asked students to choose four related words for the reloading lesson. In doing so, she demonstrated an awareness that her students develop a deeper understanding of word meanings when they explore and discuss relationships.

Ms. Blair helps students explore and use related science words with concept maps. To reload words, she provides small groups of students with a set of related terms from the word wall. She asks each group to share their words and meanings within the group, discuss their relationships, and draw arrows between the words to demonstrate those relationships. Figure 9.3 shows how students understand relationships among technical words related to erythrocyte production. The positioning of words, with the addition of arrows, expresses the relationship of biological functions being studied. Ms. Blair also asks students to express the relationships and their rationale aloud to the class.

**Figure 9.3.** Student understanding of concept relationship within erythrocyte production

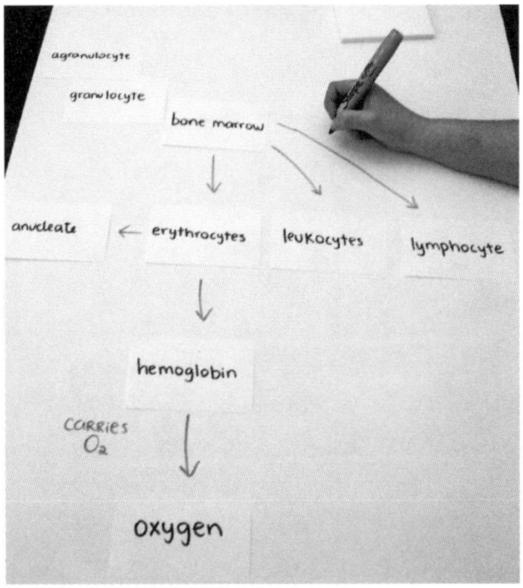

Ms. Blair then asks the groups to conduct a gallery walk. Students rotate among tables and discuss each concept map. At each table, a designated student summarizes the group's discussion. Sticky notes may be used to include comments or questions. After rotating through all of the groups, students return to their original tables and read the responses their classmates left on their concept maps. Ms. Blair concludes the reloading lesson by having students discuss similarities and differences they observed across concept maps.

She also supports EMLs' understanding that related words can be organized by meaning gradients along a continuum. She often makes use of a strategy called Shades of Meaning (adapted from Goodman 2004 as cited in Fisher and Frey 2008). For example, students use the word *accurate* in an investigation as they discuss how to display data in tables. Ms. Blair wants students to understand that there are other English synonyms for *accurate* but they do not convey the precise meaning needed in science. To visualize the meaning gradient, she asks the students to position a set of related words along a continuum using a paint chip (Figure 9.4).

For this reloading activity, Ms. Blair writes four words (*fine, correct, accurate, exactly*) on small strips of paper and places them in the middle of the table. She instructs students to discuss the meaning of each word and decide how to order them from least to most intense. To reinforce the concept that words are not just synonyms, Ms. Blair asks students to identify an additional word and develop their own meaning continuum. The student represented in Figure 9.4 demonstrated understanding by ordering the words *okay, yes!, of corse* [*of course*], and *defindly* [*definitely*].

**Figure 9.4.** Example of a Shades of Meaning activity from Ms. Blair's classroom

Center #7

okay | fine
Yes! | Correct
OFcorse | accurate
Defindly | Exactly

## Word Meanings and Usage

Another function of reloading lessons is to provide EMLs with multiple experiences with word meanings. For example, Ms. Blair makes use of a strategy called Quiz-Quiz-Trade (Kagan and Kagan 2009) to have students restate and redefine content words that were previously encountered through a science inquiry lesson (Figure 9.5). In this case, students select a word from the word wall, note its meaning, and provide examples. Using Quiz-Quiz-Trade, they repeatedly trade information with different partners as they read and discuss examples of their words. Though this type of repeated activity might be considered wasteful in classrooms in which teachers are always racing against time, it helps EMLs gain confidence as they revisit word meanings and make linguistic adjustments necessary to communicate with multiple partners.

In a different classroom, biology teacher Mr. Graham uses a four-square graphic organizer—also known as the Frayer Vocabulary Model—because it requires students to examine word meanings from various perspectives (Figure 9.6). In addition to defining a word,

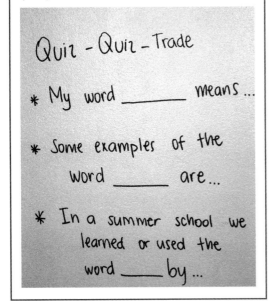

**Figure 9.5.** Scaffold provided by a teacher to help students prepare for Quiz-Quiz-Trade

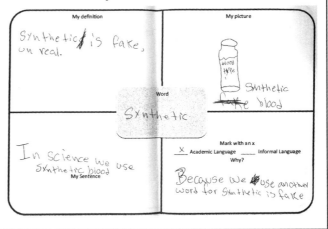

**Figure 9.6.** Four-square graphic organizer for the word *synthetic*

students consider other elements—such as characteristics and nonexamples—that support a deeper understanding of that word's meaning. Because of his interest in helping EMLs think about word usage in the context of social settings, Mr. Graham also asks students to consider whether a word is more commonly used in informal or more formal, academic settings and to provide a rationale for their response.

When introducing how to use a four-square graphic organizer, Mr. Graham leads a class discussion to highlight the four elements of the model. He finds that these group discussions help EMLs examine word meanings from different perspectives. Nevertheless, he is cautious not to overuse this strategy as students can perceive it as busywork rather than meaningful word work.

To reload word meanings, Mr. Graham also makes use of carousel feedback (Kagan and Kagan 2009). He identifies key concepts he wants students to review and writes each word on a piece of chart paper. He divides the class into small groups—three to five students per group—and assigns each group a different word and colored marker. Groups are given two minutes to discuss their assigned word and write one thing they know about it before they rotate to a new table and different chart. Using their assigned colored marker, they then add new ideas to the chart (Figure 9.7, p. 96). Once all groups have rotated to all of the charts, students return to their original tables, read the new comments, and come up with their own definition of the word or concept.

## Word Parts

Reloading lessons in your classroom can support EMLs' understanding of how word parts (morphemes) contribute to word meaning (Fang 2006; Kieffer and Lesaux 2012; Stahl and Nagy 2006). As students develop morphological awareness, they also develop the ability to

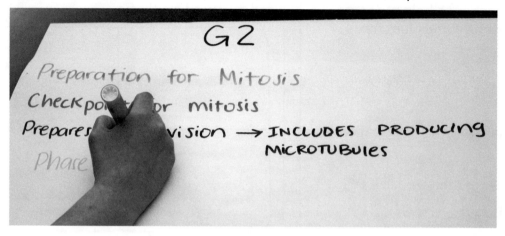

**Figure 9.7.** Carousel feedback contributions to the concept of G2

analyze word prefixes, roots, and suffixes. For example, in the opening scenario, as part of a reloading lesson, Ms. Blair helped students examine how the word *horizontal* could be broken down into two Latin-based morphemes: *horizon* + *al*. Some suffixes are more common than others. Although words ending in *-s*, *-es*, *-ed*, and *-ing* account for most of the suffixes in English (Goodwin, Lipsky, and Ahn 2012), suffixes such as *-logy*, *-phase*, and *-stasis* occur more often in science.

In their examination of how morphemes change word meanings, Ms. Blair and her students create a class chart to identify the prefixes, suffixes, and other word parts they encounter, along with a short definition for each. When teachers foster word consciousness as part of their reloading lessons, students often begin to recognize and analyze morphemes on their own as new words are introduced.

As we saw in the opening scenario, another way in which Ms. Blair taps students' understanding of morphemes is by helping Spanish speakers recognize the value of using cognates—words that have the same origin and a similar spelling and meaning in two languages—to understand English words (Freeman and Freeman 2009; Kieffer and Lesaux 2007; Quinn et al. 2012). Many Tier 2 and Tier 3 words have the same Latin and Greek roots and are cognates in Spanish. For example, the words *investigate* and *microscope* are cognates with *investigar* and *microscopio* in Spanish.

## Concluding Remarks

Reloading is used to provide EMLs with opportunities to revisit important vocabulary they encounter within the context of science lessons. Reloading makes use of a word wall to capture the important words that will be revisited throughout the lesson and unit. Reloading lessons explicitly support students in developing word consciousness by examining word relationships, parts, and meanings.

# Chapter 10

# Voices From Teachers

In writing this book, we wanted to share what we have learned while working with teachers, emergent multilingual learners (EMLs), and one another. We have conducted extensive professional development (PD) with middle and secondary teachers, and we realized that there is a broad audience who could profit from our growing knowledge. In addition to conducting PD, we have observed classes, examined recorded lessons, interviewed teachers, read journals and reflections, and conducted focus groups. These experiences have provided us with extensive feedback regarding the implementation of the 5R Instructional Model and how to support multimodal language development in science classrooms. In this chapter, we use examples from two teachers who participated in multiyear PD. On the one hand, Jackson's story describes the effects of PD sustained over four years. Audra's story, on the other hand, presents one particular instructional moment within her classroom.

## Jackson's Story

Jackson is a biology teacher who began teaching as a second career. The school into which he was placed enrolled a large number of EMLs (mainly Spanish speakers). His story points to systematic integration of the 5R Instructional Model as a framework for planning and his growing appreciation of the complexity of multimodality for the affordance of meaning-making. As is evident in his story, the process of using the 5R Instructional Model as a planning tool and understanding the hybrid language took time. Learning to teach this way is a process that does not happen immediately. You have different needs and strengths, so your journey will not be the same as Jackson's. Nevertheless, Jackson built on his prior knowledge and strengths, and we hope you will do the same.

## Year 1

Having completed his first year of teaching—and aware of the struggles he encountered—Jackson sought opportunities for professional growth. One of his colleagues shared information about a long-term PD program offered by a local university. The first year he entered our program, he wrote in an opening reflection, "I am hired to teach biology." Later, he expressed that he only needed to teach what the book and district outlined. He was most interested in pedagogical strategies for teaching content. During the content-rich summer sessions, Jackson was introduced to the 5R Instructional Model and the idea that the language of the science classroom is both social and multimodal. Observations conducted during the academic year revealed that Jackson planned and presented a good lecture with accurate biological facts. His students took notes and responded to questions, and he stressed the memorization of facts. The PD community encouraged Jackson to move from lecture to inquiry-based instruction. As he became aware of inquiry-based instruction and the benefits it provided for all students, he incorporated it more frequently. For him, a point of note was how successful his EMLs were in developing conceptual understanding in science, especially when they had repeated opportunities to encounter concepts and language. He stated, "There is a link between language and conceptual understanding that I missed before." Seeing his students' excitement, he entered the second year of the PD program eager to expand his understanding of relevant pedagogy for EMLs.

## Year 2

Jackson wanted to know more about the intersection of inquiry-based science, academic language, and the 5R Instructional Model. He also had a realistic understanding that his professional growth would take time. Aware that his biology students encounter a significant number of new terms in the classroom, he decided to focus on vocabulary. In a reflection, he wrote that he wanted "to be a resource to help my students in building vocabulary." Because the textbook provided an extensive list of vocabulary for each unit of study, Jackson started by narrowing that list to one he thought would most benefit students. For developing a greater understanding of the content, he chose essential Tier 3 words, yet he was also aware of the need to highlight general academic Tier 2 words. He purposefully selected words he thought would help students understand the specific academic tasks in which they would be engaged.

Using the 5R Instructional Model, Jackson began to plan lessons (macro-scaffolding) with a heightened awareness about how and when to introduce and review vocabulary. The *R* he considered easiest to integrate was reload. He saw the strength of reloading—as opposed to the frontloading strategies he had previously used. Jackson employed various strategies for bringing the key vocabulary back into classroom discussions. He reflected on how he had initially been afraid that if he did not front-load vocabulary, students would be unable to follow the lesson. However, once the words were introduced naturally within the context of the lesson and later reloaded, students stopped simply memorizing words—they began to make lasting connections.

As Jackson became even more comfortable working with language teaching, he decided to try to incorporate another component of the 5R Instructional Model. He realized that students did not always enter his classroom knowing scientific terms, but they did come with background knowledge about many of the topics. Considering replacement, he decided to structure his lessons in a way that invited students to engage in conversations that used terms to indicate conceptual knowledge—even if technical language was absent. During one PD session, he discussed how he had used replacement within a lesson on heredity and genetic traits. Using the students' background knowledge about animal breeding, he replaced familiar words with the academic words (e.g., *looks like = trait, baby = f1 generation*) that EMLs would need to move forward with the unit of study. In addition, he reflected, "I have the students write what is happening during the exploration stage and replace those words with scientific vocabulary." As Jackson exchanged ideas with his PD cohort members, they discussed a further benefit of using replacement. They alluded to the value of building students' background knowledge not only from an instructional perspective but also as a means to demystify science discourse.

Later in the academic year, Jackson began to plan more carefully for revealing language (macro-scaffolding). As he prepared to teach each unit, he examined the content and language to anticipate what might be replaced and what would need to be revealed. At a PD session, he discussed his plan. "I really need to think and plan ahead for what the students know and do not know. Some of you are more experienced with cellular respiration, so what words and concepts do I need to think about?"

Teachers shared their experiences, and a rich conversation ensued. Sally, a 12th-year teacher, told him that cellular respiration is a good topic for the use of visual representations. She stated that visuals help students see microscopic processes and make meaning from them. As the discussion continued, the teachers revisited the importance of helping students understand the role visual representations have in meaning-making.

## Year 3

Jackson finished his fourth year of teaching and entered his third year of PD. His first reflection for the summer was "I have taken many new ideas into my classroom. Having a better

understanding of language issues, I feel I can build bridges to help students become successful." This year, the PD used Kathy's expertise as she provided a textbook to revisit how visual representations and mathematical expressions are underused as affordances for meaning-making in the biology classroom. Through an exercise in which they examined pages in the biology text, the teachers became aware of the limitations of exclusively depending on natural language when other modes of the hybrid language afforded meaning. In the case of the textbook, the whole meaning-making affordance is greater than the sum of each mode alone.

Visual representations and mathematical expressions appear obvious to many teachers, but without intentional instruction, students may not use these modes to help them understand and communicate. A colleague in the PD commented on how helping students break down a graph provided them with an understanding of how to analyze data and use that data to explain and support a claim. Another teacher, reporting on her previous year's class, said, "Several times last year, I was fearful of focusing on multimodal language, but each time I tried, I found some success. Now I think, 'What would Cecilia, Kathy, or Molly do with this lesson?' It was painful at times but so worth it."

His peers' comments encouraged Jackson to repeat, reveal, replace, and reload mathematical expressions and visual representations. In addition, Jackson began to think about how the manual-technical operations of his inquiry-based lessons could be more than simple hands-on activities. As a teacher, he recognized that he must plan in ways that overlap the content, hybrid language, and 5R Instructional Model. This planning was necessary for him to be able to scaffold the hybrid language and incorporate the components of the 5R Instructional Model.

## Year 4

Jackson entered the summer component of the PD program expressing delight in his growing success in helping EMLs learn biology. Several teachers at his school told Jackson that students from his previous years' classes were different. They seemed to ask good questions, understand the components of an investigation, and be more willing to analyze data and construct explanations. However, as much as his students appeared to be different—and as much as he had improved as a biology teacher—he wanted his students to develop proficiency within each mode and engage in meaning-making across the modes.

During the session, he reflected that he now had a language-rich classroom but noted that something was still missing. He asked lots of questions about how he could help his students become aware of all modes of language and own them. He wanted to increase his understanding of repositioning and build further opportunities for his EMLs to engage in the discourse of science. Together, we thought more critically about what this would look and sound like in his biology classroom.

During the academic year, Jackson provided students with a number of scaffolds to help them organize and communicate meaning. He focused on fully using classroom journals so students used all modes in communicating and refining their growing knowledge. Jackson's personal PD journal entries often included reflection on students' understanding of content and learning, such as, "My students are now using many ways to communicate their understanding and argue their explanations with supporting data."

Throughout Jackson's four-year journey, we saw him initially narrow his focus by selecting components of the 5R Instructional Model and the hybrid language he wanted to implement. As he became more confident, he extended his use of both. Because of his perseverance, he was able to make instructional changes over time that ultimately benefited his students.

## Audra's Story

The second example highlights Audra, a physical science teacher with six years of experience. During a Saturday PD session, Audra shared her joy of being confident enough to change a lesson midweek as she recognized students' needs. Her growing understanding of the 5R Instructional Model and the role of multimodality to support EMLs led her to revamp her lesson in the moment (micro-scaffolding). She described a lesson in which students investigated the relationship of the height of a ball drop (potential energy) to the height of the rebound (kinetic energy). Included was a guide sheet that led students to collect mathematical data, complete a chart and graph, and answer related questions. During her PD session, Audra recounted the following.

"I began the lesson by reminding students that I had predetermined the heights from which the ball will be dropped. I asked students to identify the two variables in the investigation. They responded that they were the initial height of the drop and the rebound distance. I made the mistake of not asking students to distinguish between the manipulated and responding variables. Students worked in groups as they set up the investigation and dropped the balls. They did three trials at each height and repeatedly recorded the data in a chart. As the class ended, I assigned homework in which students were to use their graphing calculators to create a graph of their data and reconstruct it on their guide sheets. This graph would then be used to answer questions about the investigation.

"On the following day, I began by reloading the terms used the day before (e.g., *kinetic, potential, energy, variable*). My intention was to replace *manipulated variable* with *independent variable* and *responding variable* with *dependent variable*. However, during the reloading, I saw a bigger problem. I realized that the students were having trouble identifying which variable was manipulated and which was responding. This led me to realize that the graphs constructed for homework were incorrect. Some students had placed the manipulated variable on the *Y*-axis, and many had not changed the type of graphic display on the graphing calculator, so the graph they had reconstructed on their paper was a line graph rather than a scatter plot. In addition, they had not added labels to their drawings. Seeing a pattern in students' responses, I took the opportunity to discuss their prior knowledge of graphing. The discussion showed misconceptions about what is displayed by a graph and how to use a graphing calculator to help make meaning of data. I then made the decision to compile a class chart using the group data, which I had never done. This visual representation allowed me to help students see that although the measurements reported by each group were not identical, the relationships between drop and rebound were similar.

"As I planned a follow-up for this lesson, I repositioned students' understanding and language through a series of steps in which they moved from their calculators back to a pencil-and-paper graph. First, I asked students to tell me the conditions under which a line graph is appropriate. With much scaffolding, students concluded that their data did not meet the criteria (continuous data over time) and should be represented using another graphic form. Even though I am not a math teacher, I knew I had to take the opportunity to help students understand the various types of graphs and their functions.

"In addition, I guided them to think about the width of the units for the *X*- and *Y*-axes. When the class data were displayed correctly, the students were able to use the graph to construct explanations about the phenomenon they had investigated. I ended the lesson by having the students use the graphing calculator to reenter the data and create an appropriate graph. I then asked them to compare the two graphs. We discussed the reason for using a calculator and how the data could be represented by hand or by calculator output. We also talked about how, when analyzing data, each mode of representation affords meanings that best foster understanding. From the PD, I realized that this metacognition might only happen if I explicitly pointed this out to students."

With an understanding of the 5Rs, Audra was able to plan (macro-scaffold) and adjust her lesson (micro-scaffold) to seamlessly support EMLs to think and communicate using the multimodal language of science. She was able to see that students needed additional support in accomplishing her goal for the lesson; thus, she repeated, reloaded, replaced, and repositioned hybrid language in her spontaneous replanning of the lesson. As we have emphasized throughout the book, all the *R*s of the 5R Instructional Model may not occur in every lesson and may appear in any order.

## Final Remarks

We remind you that the 5R Instructional Model (replace, reveal, repeat, reposition, reload) provides a way to think about when and how we should scaffold each of the modes of the hybrid language. Membership in the community of science goes beyond proficiency in natural language and ultimately becomes about identity.

We began with the desire to help EMLs fully engage in authentic science experiences, and we ultimately developed and used the 5R Instructional Model. We purposefully called it an instructional model because we hope it will be used to plan more inclusive lessons. It is a framework for multimodal language development, not a step-by-step, linear procedure. This is what makes it both difficult and rewarding to implement. What we see as a positive is that it can be used as a way to develop macro- and micro-scaffolds to support EMLs. Because it is a planning tool, we expect you, the teacher, will see that hybrid language must be intentionally taught in the classroom.

Like Jackson and Audra, many teachers with whom we have worked have moved away from thinking that they only teach content. They have begun to understand the importance of using hybrid language as a way to support conceptual understanding and open new doors to meaning-making. We are not advocating that all teachers become English as a Second Language teachers; rather, all teachers have a responsibility to include ways of macro-scaffolding multimodal language in their planning. In addition, a full understanding of the 5R Instructional Model prepares teachers to take advantage of micro-scaffolding opportunities.

As schools become more diverse and have larger populations of EMLs, preparing teachers to address the needs of all students is imperative. This book is one step toward helping science teachers recognize and integrate multimodal language instruction into existing units. Our intent is to provide an instructional model teachers can use over time.

In our own work, we know the collaboration that has intersected our disciplines has been critical to our growth as teachers and researchers. The collaboration that started with the need to design a summer program for newcomers transformed our teaching practices. We recognize that, alone, none of us would have arrived at our current understanding of the complexity of the discourse of the science community, but as a collaborative team, we provide insight to one another. We hope you find the opportunity to work with others in providing the best instruction for EMLs to be as challenging, stimulating, and rewarding as we have.

# References

Abrahams, I., and R. Millar. 2008. Does practical work really work? A study of the effectiveness of practical work as a teaching and learning method in school science. *International Journal of Science Education* 30 (14): 1945–1969.

Ainsworth, S., V. Prain, and R. Tytler. 2011. Drawing to learn in science. *Science* 333 (6046): 1096–1097.

American Association for the Advancement of Science (AAAS). 1993. *Benchmarks for science literacy.* New York: Oxford University Press.

Anderson, R. C., and W. E. Nagy. 1993. *The vocabulary conundrum* (Center for the Study of Reading Technical Report No. 570). Urbana-Champaign, IL: University of Illinois, Urbana-Champaign.

Anstey, M., and G. Bull. 2006. *Teaching and learning multiliteracies: Changing times, changing literacies.* Newark, DE: International Reading Association.

Beck, I. L., and M. G. McKeown. 1985. Teaching vocabulary: Making the instruction fit the goal. *Educational Perspectives* 23 (1): 11–15.

Bezemer, J., and G. Kress. 2016. *Mutimodality, learning and communication.* New York: Routledge.

Blachowicz, C., and P. J. Fisher. 2006. *Teaching vocabulary in all classrooms.* Upper Saddle River, NJ: Pearson.

Brisk, M.A., and Q. Zhang-Wu. 2017. Academic language in K–12 contexts. In *Handbook of research in second language teaching and learning*, ed. E. Hinkel, 82. New York: Routledge.

Bunch, G. C., J. M. Shaw, and E. R. Geaney. 2010. Documenting the language demands of mainstream content-area assessment for English learners: Participant structures, communicative modes and genre in science performance assessments. *Language and Education* 24 (3): 185–214.

Bybee, R., J. A. Taylor, A. Gardner, P. Van Scotter, J. C. Powell, A. Westbrook, and N. Landes. 2006. *The BSCS 5E instructional model: Origins, effectiveness, and applications.* Colorado Springs, CO: BSCS.

Cajori, F. 1928. *A history of mathematical notations.* London, UK: Open Court.

Carlo, M. S., D. August, B. McLaughlin, C. E. Snow, C. Dressler, D. N. Lippman, T. J. Lively, and C. E. White. 2004. Closing the gap: Addressing the vocabulary needs of English language learners in bilingual and mainstream classrooms. *Reading Research Quarterly* 39 (2): 188–206.

Cartier, J. L, M. Smith, M. K. Stein, and D. K. Ross. 2013. *5 Practices for orchestrating productive task-based discussions in science.* Arlington, VA: NSTM & NSTA.

Cazden, C. B. 1988. *Classroom discourse. The language of teaching and learning.* Portsmouth, NH: Heinemann.

Chamot, A. U. 2005. Language learning strategy instruction: Current issues and research. *Annual Review of Applied Linguistics* 25 (1): 112–130.

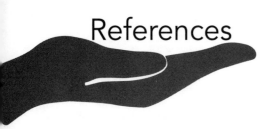

# References

Chamot, A. U., and J. M. O'Malley. 1994. *The CALLA handbook*. Reading, MA: Addison-Wesley.

Cole, A. 2006. Scaffolding beginning readers: Micro and macro cues teachers use during student oral reading. *The Reading Teacher* 59 (5): 450–459.

Coleman, J. M., L. G. Bradley, and C. A. Donovan. 2012. Visual representations in second graders' information book compositions. *The Reading Teacher* 66 (1): 31–45.

Coleman, J. M., E. M. McTigue, and L. B. Smolkin. 2011. Elementary teachers' use of graphical representations in science teaching. *Journal of Science Teacher Education* 22 (7): 613–643.

Collier, V. 1989. How long? A synthesis of research in academic achievement in a second language. *TESOL Quarterly* 23: 509–531.

Cummins, J. 1981. *Schooling and language minority students: A theoretical framework*. Los Angeles: Evaluation, Dissemination, and Assessment Center, California State University.

Cummins, J. 1994. Knowledge, power, and identity in teaching English as a second language. In *Educating second language children: The whole child, the whole curriculum, the whole community*, ed. F. Genesse, 33. New York: Cambridge University Press.

Cummins, J. 1996. *Negotiating identities: Education for empowerment in diverse society*. Los Angeles: California Association for Bilingual Education.

Cummins, J. 2000. *Language, power and pedagogy: Bilingual children in the crossfire*. Clevedon, UK: Multilingual Matters.

Daniels, H., A. Edwards, Y. Engestrom, T. Gallager, and S. R. Ludvigsen, eds. 2010. *Activity theory in practice. Promoting learning across boundaries and agencies*. London: Routledge.

de Oliveira, L. C. 2017. A language-based approach to content instruction (LACI) in science for English language learners. In *Science teacher preparation in content-based second language acquisition*, eds. A. W. Oliveira and M. H. Weinburgh, 41. Switzerland: Springer International.

de Oliveira, L. C., and S. W. Lan. 2014. Writing science in an upper elementary classroom: A genre-based approach to teaching English language learners. *Journal of Second Language Writing* 25 (1): 23–39.

de Jong, E. J., C. A. Harper, and M. R. Coady. 2013. Enhanced knowledge and skills for elementary mainstream teachers of English language learners. *Theory into Practice* 52 (2): 89–97.

Díaz-Rico, L. T., and K. Z. Weed. 2010. *The cross-cultural language and academic development handbook*. 4th ed. Boston: Allyn & Bacon.

DiCerbo, P. A., K. A. Anstrom, L. L. Baker, and C. Rivera. 2014. A review of the literature on teaching academic English to English language learners. *Review of Educational Research* 84 (3): 446–482.

Donovan, C. A., and L. B. Smolkin. 2001. Genre and other factors influencing teachers' book selections for science instruction. *Reading Research Quarterly* 36 (4): 412–440.

Echevarría, J., M. Vogt, and D. J. Short. (2000). *The SIOP model for teaching mathematics to English learners*. Boston: Pearson.

Faggella-Luby, M., R. Griffith, C. Silva, and M. H. Weinburgh. 2016. Assessing ELLs' reading comprehension and science language development. *Electronic Journal of Science Education* 20 (3): 150–166.

Fang, Z. 2006. The language demands of science reading in middle school. *International Journal of Science Education* 28 (5): 491–520.

Fang, Z. 2013. Disciplinary literacy in science. *Journal of Adolescent & Adult Literacy*, 57 (4): 274–278.

Fang, Z., L. L. Lamme, and R. M. Pringle. 2010. *Language and literacy in inquiry-based science classroom grades 3–8*. Arlington, VA: NSTA Press.

Fang, Z., and M. Schleppegrell. 2008. *Reading in secondary content areas: A language-based pedagogy*. Ann Arbor, MI: University of Michigan Press.

Fathman, A. K., and D. T. Crowther, eds. 2006. *Science for English language learners: K–12 classroom strategies*. Arlington, VA: NSTA Press.

Fisher, D., and N. Frey. 2008. *Word wise and content rich*. Portsmouth, NH: Heinemann.

Ford, D. J. 2004. High recommended trade books: Can they be used in inquiry science? In *Border crossings: Essays on literacy and science*, ed. E. W. Saud, 277. Arlington, VA: NSTA Press and Newark, DE: International Reading Association.

Ford, D. J. 2006. Representations of science within children's trade books. *Journal of Research in Science Teaching* 43 (2): 214–235.

Freeman, Y. S., and D. E. Freeman. 2009. *Academic language for English learners and struggling readers*. Portsmouth, NH: Heinemann.

Freire, P. 1970. *Pedagogy of the oppressed*. New York: Continuum.

Fries-Gaither, J., and T. Shiverdecker. 2012. *Inquiring scientists, inquiring readers*. Arlington, VA: NSTA Press.

Gee, J. P. 2000. Teenagers in new times: A new literacy studies perspective. *Journal of Adolescent & Adult Literacy* 43 (5): 412–420.

Gee, J. P. 2002. Literacies, identities, and discourses. In *Developing advanced literacy in first and second languages*, ed. M. J. Schleppegrell and M. C. Colombi, 159–175. Mahwah, NJ: Lawrence Erlbaum.

Gee, J. P. 2004. Language in the science classroom: Academic social languages as the heart of school-based literacy. In *Crossing borders in literacy and science instruction*, ed. E. W. Saul, 13. Arlington, VA: NSTA Press.

Gee, J. P. 2008. What is academic language? In *Teaching science to English language learners*, ed. A. S. Rosebery and B. Waren, 57. Arlington, VA: NSTA Press.

Gee, J. P. 2014. Decontextualized language: A problem, not a solution. *International Multilingual Research Journal* 8 (1), 9–23.

Gibbons, P. 2002. *Scaffolding language, scaffolding learning: Teaching second language learners in the mainstream classroom*. Portsmouth, NH: Heinemann.

Gibbons, P. 2003. Mediating language learning: Teacher interactions with ESL students in a content-based classroom. *TESOL Quarterly* 37 (2): 247–273.

Gibbons, P. 2006. *Bridging discourses in the ESL classroom*. London & New York: Continuum.

Gibbons, P. 2009. *English learners, academic literacy, and thinking*. Portsmouth, NH: Heinemann.

Gibbons, P. 2015. *Scaffolding language, scaffolding learning*. 2nd ed. Portsmouth, NH: Heinemann.

Goodwin, A., M. Lipsky, and S. Ahn. 2012. Word detectives: Using units of meaning to support literacy. *Reading Teacher* 65 (7): 461–470.

Graves, M. F., D. August, and J. Mancilla-Martinez. 2013. *Teaching vocabulary to English language learners*. New York: Teachers College Press, Center for Applied Linguistics, International Reading Association, Teachers of English to Speakers of Other Languages.

Groves, F. H. 1995. Science vocabulary load of selected secondary science textbooks. *School Science and Mathematics* 95 (5): 231–235.

Gunel, M., and F. Yesildag-Hasancebi. 2016. Modal representations and their role in the learning process: A theoretical and pragmatic analysis.

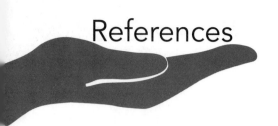

# References

*Educational Sciences: Theory & Practice* 16 (1): 109–126.

Hakuta, K., Y. Butler, and D. Witt. 2000. *How long does it take for English learners to attain proficiency?* Santa Barbara, CA: University of California, Linguistic Minorities Project.

Halliday, M. 1993. Towards a language-base theory of learning. *Linguistics and Education* 5: 93–116.

Hammond, J. 2014. An Australian perspective on standards-based education, teacher knowledge, and students of English as an additional language. *TESOL Quarterly* 48 (3): 507–532.

Hammond, J., and P. Gibbons. 2005. Putting scaffolding to work: The contributions of scaffolding in articulating ESL education. *Prospect* 20 (1): 6–30.

Hand, B. M. 2008. Introducing the science writing heuristic approach. In *Science inquiry, argument, and language*, ed. B. M. Hand, 1. Rotterdam, the Netherlands: Sense.

Hand, B., and C. Keys. 1999. Inquiry investigation: A new approach to laboratory reports. *Science Teacher* 66 (4): 27–29.

Hand, B., M. McDermott, and V. Prain, eds. 2016. *Using multimodal representations to support learning in the science classrooms.* New York: Springer.

Hattie, J., and G. Yates. 2014. *Visible learning and the science of how we learn.* New York: Routledge.

Haug, B. S., and M. Ødegaard. 2014. From words to concepts: Focusing on word knowledge when teaching for conceptual understanding within an inquiry-based science setting. *Research in Science Education* 44: 777–800.

Heritage, M., A. Walqui, and R. Linquanti. 2015. *English language learners and the new standards.* Cambridge, MA: Harvard Education.

Hunter, A. B., S. L. Laursen, and E. Seymour. 2007. Becoming a scientist: The role of undergraduate research in students' cognitive, personal, &

professional development. *Science Education* 91 (1): 36–74.

Jewitt, C., ed. 2017. *The Routledge handbook of multimodal analysis.* New York: Routledge.

Jornet, A., and W.-M. Roth. 2015. The joint work of connecting multiple (re)presentations in science classrooms. *Science Education* 99 (2): 378–403.

Kagan, S. 1995. We can talk: Cooperative learning in the elementary ESL classroom. *Kagan Online Magazine,* Spring. San Clemente, CA: Kagan Publishing.

Kagan, S., and M. Kagan. 2009. *Cooperative learning.* San Juan Capistrano, CA: Kagan Cooperative Learning.

Karplus, R., and H. D. Thier. 1967. *A new look at elementary school science.* Chicago: Rand McNally.

Keys, C. W., B. Hand, V. Prain, and S. Collins. 1999. Using the science writing heuristic as a tool for learning laboratory investigations in secondary science. *Journal of Research in Science Teaching* 36 (10): 1065–1084.

Kieffer, M. J., and N. K. Lesaux. 2007. Breaking down words to build meaning: Morphology, vocabulary, and reading comprehension in the urban classroom. *The Reading Teacher* 61 (2): 134–144.

Kieffer, M. J., and N. K. Lesaux. 2012. Effects of academic language instruction on relational and syntactic aspects of morphological awareness for sixth graders from linguistically diverse backgrounds. *Elementary School Journal* 112 (3): 519–545.

Krashen, S. D. 2003. *Explorations in language acquisition and use.* Portsmouth, NH: Heinemann.

Kress, G. 2010. *Multimodality: A social semiotic approach to contemporary communication.* London: Routledge.

Kress, G., C. Jewitt, J. Ogborn, and C. Tsatsarelis. 2001. *Multimodal teaching and learning: The rhetorics of the science classroom.* London & New York: Continuum.

Kucer, S. B., and C. Silva. 2013. *Teaching the dimensions of literacy*. New York: Routledge.

Kucer, S. B., C. Silva, and E. L. Delgado-Larocco. 1995. *Curricular conversations: Themes in multilingual and monolingual classrooms*. York, MA: Stenhouse.

Kutz, E. 1997. *Language and literacy: Studying discourse in communities and classrooms*. Portsmouth, NH: Heinemann.

Larsen-Freeman, D. 2012. On the roles of repetition in language teaching and learning. *Applied Linguistics Review* 3 (2): 195–210.

Larsen-Freeman, D. 2014. Let learning emerge. *Language Magazine* 13 (7): 24–27.

Lave, J., and E. Wenger. 1991. *Situated learning: Legitimate peripheral participation*. New York: Cambridge University Press.

Lee, O. 2005. Science education with English language learners: Synthesis and research agenda. *Review of Educational Research* 75 (4): 491–530.

Lee, O., E. Miller, and R. Januszyk. 2015. NGSS *for all students*. Arlington, VA: NSTA Press.

Lee, O., H. Quinn, and G. Valdéz, G. 2013. Science and language for English language learners in relation to the Next Generation Science Standards and with implications for Common Core State Standards for English language arts and mathematics. *Educational Researcher* 42 (4): 223–233.

Lemke, J. L. 1989. Making text talk. *Theory into practice* 28 (2): 136–141.

Lemke, J. L. 1990. *Talking science*. Westport, CT: Ablex.

Lemke, J. L. 1998a. Teaching all languages of science: Words, symbols, images, and actions. Brooklyn College, City University of New York. *http://academic.brooklyn.cuny.edu/education/jlemke/papers/barcelon.htm*.

Lemke, J. L. 1998b. Multimedia literacy demands on the scientific curriculum. *Linguistics and Education* 10 (3): 247–271.

Lemke, J. L. 2002. Multimedia semiotics: Genres for science education and scientific literacy. In *Developing advanced literacy in first and second languages*, ed. M. J. Schleppegrell and M. C. Colombi, 21–44. Mahwah, NJ: Lawrence Erlbaum.

Lemke, J. L. 2004. The literacies of science. In *Crossing borders in literacy and science instruction*, ed. E.W. Saul, 33–67. Arlington, VA: NSTA Press.

Livers, S. D., and J. M. Bay-Williams. 2014. Vocabulary support: Constructing (not obstructing) meaning. *Mathematics Teaching in the Middle School* 20 (3): 152–159.

Luft, J., R. L. Bell, and J. Gess-Newsome, eds. 2008. *Science as inquiry in the secondary setting*. Arlington, VA: NSTA Press.

Macaro, E. 2006. Strategies for language learning and for language use: Revising the theoretical framework. *The Modern Language Journal* 90 (3): 320–337.

Martínez, R. A., M. F. Orellana, M. Pacheco, and P. Carbone. 2008. Found in translation: Connecting translating experiences to academic writing. *Language Arts* 85 (6): 421–431.

McTigue, E. M., and A. C. Flowers. 2011. Science visual literacy: Learners' perceptions and knowledge of diagrams. *The Reading Teacher* 64 (8): 578–589.

Met, M. 1994. Teaching content through a second language. In *Educating second language children: The whole child, the whole curriculum, the whole community*, ed. F. Genesee, 159. Cambridge, UK: Cambridge University Press.

Moje, E. B. 2007. Developing socially just subject-matter instruction: A review of the literature on disciplinary literacy teaching. *Review of Research in Education* 31 (1): 1–44.

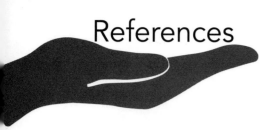

# References

Moje, E. B, K. M. Ciechanowski, K. Kramer, L. Ellis, R. Carrillo, and T. Collazo. 2004. Working toward third space in content area literacy: An examination of everyday funds of knowledge and discourse. *Reading Research Quarterly* 39 (1): 38–70.

Moline, S. 2012. *I see what you mean.* Portland, ME: Stenhouse.

Molle, D. 2015. Academic language and academic literacies: Mapping a relationship. In *Multilingual learners and academic literacies*, ed. D. Molle, E. Sato, T. Boals, and C. A. Hedgspeth, 13. New York: Routledge.

Molle, D., E. Sato, T. Boals, and C. A. Hedgspeth, eds. 2015. *Multilingual learners and academic literacies.* New York: Taylor & Francis.

Murcia, K. 2010. Multi-modal representations in primary science: What's offered by interactive whiteboard technology. *Teaching Science* 56 (1): 23–29.

National Council of Teachers of Mathematics (NCTA). 2014. *Principles to actions: Ensuring mathematical success for all.* Reston, VA: NCTA.

National Research Council. 1996. *National science education standards.* Washington, DC: National Academies Press.

National Research Council. 2000. *Inquiry and the national science education standards.* Washington, DC: National Academies Press.

National Research Council. 2012. *A framework for K–12 science education: Practices, crosscutting concepts, and core ideas.* Washington, DC: National Academies Press.

NGSS Lead States. 2013. *Next Generation Science Standards: For states, by states.* Washington, DC: National Academies Press.

Oliveira, A. W., and M. H. Weinburgh, eds. 2017. *Science teacher preparation in content-based second language acquisition.* Dordrecht, the Netherlands: Springer.

Orellana, M. F., and J. F. Reynolds. 2008. Cultural modeling: Leveraging bilingual skills for school paraphrasing tasks. *Reading Research Quarterly* 43 (1): 48–65.

Oxford, R. L. 1990. *Language learning strategies.* Boston: Heinle.

Pappas, C. 2006. The information book genre: Its role in integrated science literacy research and practice. *Reading Research Quarterly* 41 (2): 226–250.

Pearn, C. A. 2007. Using paper folding, fraction walls, and number lines to develop understanding of fraction with students from years 5–8. *Australian Mathematics Teacher* 63 (4): 31–36.

Piaget, J. 1951. *Psychology of intelligence.* London: Routledge and Kegan Paul.

Piaget, J. 1953. *Logic and psychology.* Manchester, UK: Manchester University Press.

Polanyi, M. 1967. *The tacit dimension.* New York: Anchor.

Polias, J. 2015. *Apprenticing students into science: Doing, talking, writing, and drawing scientifically.* Stockholm, Sweden: Hallgreen and Fallgren.

Prain, V., and R. Tytler. 2013. Representing and learning in science. In *Constructing Representations to Learn in Science*, ed. R. Tytler, V. Prain, P. Hubber, and B. Waldrip, 1–14. Rotterdam, the Netherlands: Sense.

Prain, V., and R. Waldrip. 2006. An exploratory study of teachers' and students' use of multi-modal representations of concepts in primary science. *International Journal of Science Education* 28 (15): 1843–1866.

Quinn, H., O. Lee, and G. Valdes. 2012. *Language demands and opportunities in relation to Next Generation Science Standards for English language learners: What teachers need to know* (Understanding

Language Initiative). Stanford, CA: Stanford University.

Reeves, J. 2006. Secondary teacher attitudes toward including English-language learners in mainstream classrooms. *The Journal of Educational Research* 99 (3): 131–42.

Richmond, G., and L. A. Kurth. 1999. Moving from outside to inside: High school students' use of apprenticeships as vehicles for entering the culture and practice of science. *Journal of Research in Science Teaching* 36: 677–697.

Roessingh, H. 2005. The intentional teacher: The mediator of meaning making. *The Journal of Educational Though (JET)/Revue de la Pensée Éducative* 39 (2): 111–134.

Rothenberg, C., and D. Fisher. 2007. *Teaching English language learners, a differentiated approach.* Columbus, OH: Pearson.

Sadler, T. D., S. Burgin, L. McKinney, and L. Ponjuan. 2010. Learning science through research apprenticeships: A critical review of the literature. *Journal of Research in Science Teaching* 47 (3): 235–256.

Saul, E. W., ed. 2004. *Crossing boarders in literacy and science instruction.* Arlington, VA: NSTA Press.

Saul, E. W., and D. Dieckman. 2005. Theory and research into practice: Choosing and using information trade books. *Reading Research Quarterly* 40 (4): 502–513.

Scarcella, R. 2003. Academic English: A conceptual framework. [The University of California Linguistic Minority Research Institute: Technical Report 20003-1]. *www.lmri.ucsb.edu.*

Schleppegrell, M. J. 2004. *The language of schooling: A functional linguistic perspective.* Mahwah, NJ: Lawrence Erlbaum.

Scieszka, J., and L. Smith. 1995. *Math curse.* New York, NY: Viking.

Scott, J. A. 2005. Creating opportunities to acquire new word meanings from text. In *Teaching and learning vocabulary: Bridging research to practice*, ed. E. H. Hiebert and M. L. Kamil. Mahwah, NJ: Lawrence Erlbaum.

Scott, J. A., and W. E. Nagy. 2009. Developing word consciousness. In *Essential readings on vocabulary instruction*, ed. M. F. Graves, 102. Newark, DE: International Reading Association.

Serafini, F. 2011. Expanding perspectives for comprehending visual images in multimodal texts. *Journal of Adolescent & Adult Literacy* 54 (February): 342–350.

Settlage, J. 2007. Demythologizing science teacher education: Conquering the false ideal of open inquiry. *Journal of Science Teacher Education* 18 (4): 461–467.

Settlage, J., A. Madsen, and K. Rustad. 2005. Inquiry science, sheltered instruction, and English language learners: Conflicting pedagogies in highly diverse classrooms. *Issues in Teacher Education* 14 (1): 39–57.

Short, K. G., J. C. Harste, and C. L. Burke. 1996. *Creating classrooms for authors and inquirers.* Portsmouth, NH: Heinemann.

Silva, C., M. H. Weinburgh, and K. H. Smith. 2013. Not just good science teaching: Supporting academic language development. *Voices in the Middle* 20 (3): 34–42

Silva, C., M. H. Weinburgh, K. H. Smith, G. Barreto, and J. Gabel. 2009. Partnering to develop academic language for English language learners through mathematics and science. *Childhood Education* 85 (2): 107–112.

Silva, C., M. H. Weinburgh, K. Smith, R. Malloy, and J. Marshall (Nettles). 2012. Toward integration: A model of science and literacy. *Childhood Education* 88 (2): 91–95.

Slavin, R. 1995. *Cooperative learning: Theory, research, and practice*, 2nd ed. Boston: Allyn & Bacon.

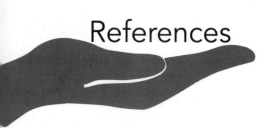

# References

Snow, C. E., and P. Uccelli. 2009. The challenges of academic language. In *The Cambridge handbook of literacy*, ed. R. D. Olson and N. Torrance, 112–133. New York: Cambridge University Press.

Stahl, S. A., and W. E. Nagy. 2006. *Teaching word meanings*. Mahwah, NJ: Lawrence Erlbaum.

Sundby, S. 2000. *Cut down to size at high noon*. Watertown, MA: Charlesbridge.

Swain, M. 1985. Communicative competence: Some roles of comprehensible input and comprehensible output in its development. In *Input in second language acquisition*, ed. S. Gass and C. Madden, 235. New York: Newbury House.

Swain, M. 2005. The output hypothesis. In *Handbook of research in second language teaching and learning*, ed. E. Hinkel, 471. Mahwah, NJ: Lawrence Erlbaum.

TESOL International Association. 2006. *PreK–12 English language proficiency standards.* Alexandria, VA: TESOL.

Treagust, D. R. 2007. General instructional methods and strategies. In *Handbook of Research on Science Education*, ed. S. Abel and N. Lederman, 373. Mahwah, NJ: Lawrence Erlbaum.

Veel, R. 1997. Learning how to mean—scientifically speaking: Apprenticeship into scientific discourse in the secondary school. In *Genre and institutions: Social processes in the workplace and school*, ed. F. Christie and J. R. Martin, 161–195. London & New York: Continuum.

Vygotsky, L. S. 1968. *Thought and language*. Cambridge, MA: MIT Press.

Vygotsky, L. S. 1978. *The mind in society*. Cambridge MA: Harvard University Press.

Walqui, A. 2006. Scaffolding instruction for English language learners: A conceptual framework. *The International Journal of Bilingual Education and Bilingualism* 9 (2): 159–180.

Warren, B., and A. S. Rosebery. (2008). Essay: Using everyday experience to teach science. In *Teaching science to English language learners*, ed. A. Rosebery and B. Warren. Arlington, VA: NSTA press.

Wee, B., D. Shepardson, J. Fast, and U. J. Harbor. 2007. Teaching and learning about inquiry: Insights and challenges in professional development. *Journal of Science Techer Education* 18 (1): 63–89.

Weinburgh, M. H., and C. Silva. 2011a. Integrating language and science: The 5Rs for English language learners. In *Science and Mathematics: International Innovations, Research, and Practices*, ed. D. F. Berlin and A. L. White, 19. Columbus, OH: International Consortium for Research in Science and Mathematics Education.

Weinburgh, M. H., and C. Silva. 2011b. Math, science, and models. *Science & Children* 48 (10): 38–42.

Weinburgh, M. H., and C. Silva. 2013. An instructional theory for English language learners: The 5R model for enhancing academic language development in inquiry-based science. In *The handbook of educational theories*, ed. B. J. Irby, G. Brown, and R. Lara-Alecio, and sect. ed. J. Koch, 293–301. Charlotte, NC: Information Age Publishing.

Weinburgh, M. H., C. Silva, and K. H. Smith. 2014. Is this a science, mathematics, or language arts lesson? Practical advice for teachers of students learning English. In *Initiatives in Mathematics and Science Education with Global Implications*, ed. D. F. Berlin and A. L. White, 93. Columbus, OH: International Consortium for Research in Science and Mathematics Education.

Weinburgh, M. H., C. Silva, R. Malloy, J. Marshall, and K. Smith. 2012. A science lesson or language lesson? Using the 5Rs. *Science & Children* 49 (9): 72–76.

Weinburgh, M., S. Silva, K. Smith, J. Groulx, and J. Nettles. 2014. The intersection of inquiry-based science and language: Preparing teachers for ELL

classrooms. *Journal of Science Teacher Education* 25 (5): 519–542.

West, S. S. 2010. An analysis of the descriptive, comparative and experimental scientific research designs in the 2009 Texas Essential Knowledge and Skills (TEKS). *The Texas Science Teacher* 39 (1): 20–29.

Wright, L. 2015. Inquiry to acquire: A discourse analysis of bilingual students' development of science literacy. In *Multilingual learners and academic literacies*, ed. D. Molle, E. Sato, T. Boals, and C. A. Hedgspeth. New York: Taylor & Francis.

Yager, R. E., H. Akcay, A. Choi, and S. O. Yager. 2009. Student success in recognizing definition of eight terms found in fourth grade science textbooks. *Electronic Journal of Science Education* 13 (2): 83–99.

Zarnowski, M., and S. Turkel. 2013. How nonfiction reveals the nature of science. *Children's Literature in Education 44* (4): 295–310.

Zhang, W. 2017. A functional approach to twenty-first century science literacy. In *Science teacher preparation in content-based second language acquisition,* ed. A. W. Oliveira and M. H. Weinburgh, 271. Switzerland: Springer International.

Zhang, Y. 2016. Multimodal teacher input and science learning in a middle school sheltered classroom. *Journal of Research in Science Teaching* 53 (1): 7–30.

# Index

Page numbers printed in **boldface type** refer to figures or tables.

# Index

integrated physics and chemistry class example, 45–46, **45**, **46**, 48
manual-technical operations, 54, **55**
mathematical expressions, 50–53, **51**
natural language, 48, **49**, 50
rationale, 46–47
visual representations, 53–54, **53**
reposition
described, 37, **41**, 42, 88
energy and physical science class example, 75–76, **76**
erosion class example, 87, **87**
manual-technical operations, 87, **87**
mathematical expressions, 82–85, **83**, **84**, **85**
natural language, 78–82, **78**, **79**, **80**, **81**
rationale, 77
visual representations, 85–86, **85**, **86**
reveal
described, 37, **41**, 42, 65
manual-technical operations, 64–65
mathematical expressions, 61–63, **62**, **63**
natural language, 59–61
physics class example, 57–58, 59–60
rationale, 58–59
visual representations, 63–64, **64**
scaffolding, 37, 38–41, 46, 51, 53, 58–59, 60, 72, 77, **95**, 100, 102–103
*See also* crime scene investigation (CSI) unit example; erosion class example
front-loading vocabulary, 58–59

## H

hybrid language of science classroom
CSI unit example, 9–10, **10**, 11–12, 16, 19–20, **20**
manual-technical operations, 9, 18, **19**, **20**
mathematical expressions, 9, 15–16, **16**, **20**
meaning-making support, 13
natural language, 9, 13–15, **15**, **20**
rationale, 10–11
social language of science, 11–12
visual representations, 9, 16–18, **17**, **20**

## I

inquiry-based science instruction
as context for communication, 22–24, **24**
described, 25–26
erosion class example, 32–34, **32**, **33**
inquiry continuum, 26–29, **27–28**
investigation types, 28–29
and language integration, 7–8
multimodality of science, 24–25
PD for content knowledge and language development, 21–22
practices of science, 29, **30–31**
interaction and macro-scaffolding, 40
iteration, 69

## L

learning and language
construction model, 5, 6–7
inquiry-based science and language integration, 7–8
scaffolding, 38–41
transmission model, 5–6
*See also* hybrid language of science classroom; words (reload)

## M

macro-scaffolding, 39–40, 48, 60, 69, 77, 92, 98–99, 102, 103
manual-technical operations, 9, 18, **19**, **20**
repeat, 72–73, **73**
replace, 54, **55**
reposition, 87, **87**
reveal, 64–65
mathematical expressions, 9, 15–16, **16**, **20**
repeat, 70–71
replace, 50–53, **51**
reposition, 82–85, **83**, **84**, **85**
reveal, 61–63, **62**, **63**
mathematics, using mathematics and computational thinking, **31**
message abundancy, 69
metacognitive awareness and macro-scaffolding, 39

# Index